Ecology and Evolution in Anoxic Worlds

Oxford Series in Ecology and Evolution
Edited by Robert M. May and Paul H. Harvey

Ecology and Evolution in Anoxic Worlds

Tom Fenchel

Marine Biological Laboratory
University of Copenhagen

Bland J. Finlay

NERC Institute of Freshwater Ecology
Windermere Laboratory

Oxford New York Tokyo
OXFORD UNIVERSITY PRESS
1995

Oxford University Press, Walton Street, Oxford OX2 6DP
Oxford   New York
Athens   Auckland   Bangkok   Bombay
Calcutta   Cape town   Dares Salaam   Delhi
Florence   Hong Kong   Istanbul   Karachi
Kuala lumpur   Madras   Madrid   Melbourne
Mexico City   Nairobi   Paris   Singapore
Taipei   Tokyo   Toronto

and associated companies in
Berlin   Ibadan

Oxford is a trade mark of Oxford University Press

Published in the United States
by Oxford University Press Inc., New York

© Tom Fenchel and Bland J. Finlay, 1995

A catalogue record for this book is available from the British Library

Library of Congress Cataloging in Publication Data
Fenchel, Tom.
Ecology and evolution in anoxic worlds/Tom Fenchel, Bland J. Finlay.
(Oxford series in ecology and evolution)
Includes bibliographical references and indexes.
1. Anaerobiosis.   I. Finlay, Bland J.   II. Title.   III. Series.
QH518.5.F46   1995   574.1'28—dc20   94-35058

ISBN   0 19 854838 9 (Hbk)
ISBN   0 19 854837 0 (Pbk)

Typeset by AMA Graphics Ltd., Preston, Lancs
Printed in Great Britain on acid free paper by
Redwood Books, Trowbridge

Acknowledgements

Much of our work during the last five years, some of which forms the basis of this book, has been based on support from various sources. TF acknowledges grants from the Danish Natural Science Research Council (nos 11-8391 and 0088-1) and from the European Union (MAST-Programme, nos 0044-C and 93-0058). BJF acknowledges support from the Natural Environment Research Council (UK).

Contents

Introduction

This is a book about the natural history of anoxic environments and their microbial inhabitants. Many people believe that oxygen is an absolute prerequisite for life. This is true for all plants, higher fungi, and almost all animals (metazoa), and it is also true for many prokaryotic and eukaryotic microorganisms. But many prokaryotes, and some unicellular eukaryotes, thrive in the absence of oxygen and in many cases they are even inhibited or killed by trace amounts of this element. Life originated under anaerobic conditions, and atmospheric oxygen is a result of biological evolution. Oxygenic photosynthesis appeared very early in the evolution of life (earlier than appreciated only a few years ago), but most of the fundamental biochemistry and metabolic pathways had evolved before the advent of atmospheric oxygen.

Anaerobic organisms are not neglected in the literature, mainly because they impinge upon a variety of economically important problems and processes (e.g. fermented foods and beverages, parasites of medical and veterinary importance, soil fertility, corrosion of iron, sewage treatment and biological methane production). During the last century, and especially during the last two decades, microbiologists have discovered many new types of anaerobic prokaryotes, and they have unravelled their energy metabolism and other physiological properties. Much of this work, which was recently reviewed by Zehnder (1988), also includes ecological insight, especially with respect to interactions between anaerobic bacteria and how the mechanisms of interaction may explain the diversity of anaerobic communities. It has been appreciated for some time that many important biogeochemical processes in the geological past as well as in the contemporary biosphere take place only under anaerobic conditions.

So, why write another book about anaerobic life? Although some limited aspects of anaerobiosis and particular anaerobic habitats have been reviewed recently, a synthesis of the ecology and evolution of anaerobic habitats and their biota has obviously been lacking. Superficially, anaerobic communities as disparate as those in the hindgut of termites and in the deeper waters of the Black Sea, may seem to have little in common. But they do, in fact, share a number of important properties which separate them from the biotic communities of the oxic world. Anaerobic biota consist almost exclusively of pro-and eukaryotic microbes. Characteristically, each species has a limited metabolic repertoire, and the species diversity of the anaerobic community is in reality a reflection of the diversity in energy metabolism. Mineralization of organic substrates is therefore the result of the concerted activity of different physiological types of organisms. Syntrophy (i.e. metabolites of one species serving as substrates for other species) and competition for substrates are the principal interspecific interactions which determine the structure of anaerobic communities. Phagotrophy (i.e. predation), which is the main type of interaction in aerobic communities, plays a much smaller role in anaerobic communities. The reason for this is that growth yields associated with anaerobic metabolism are low; so food chains rarely involve more than one trophic level of phagotrophs. The importance of molecular diffusion in the transport of substrates and metabolites is another unusual feature which strongly influences the physical structure of microbial communities.

Much progress has recently been made in our understanding of early (Precambrian) life. Notable events include the discovery of Precambrian prokaryote and eukaryote fossils, the interpretation of stromatolites and the study of their contemporary counterparts (cyanobacterial mats), new geological evidence for the increasing concentration of atmospheric oxygen during the Proterozoic, the understanding of the evolutionary derivation of eukaryote organelles from endosymbiotic bacteria, and, not least, fresh and radically different insight into the phylogeny of microorganisms, obtained from analysing rRNA-gene sequences.

It is tempting to use contemporary anaerobic life as models for early Precambrian communities and we consider this in some detail. In some respects, contemporary anaerobic microbial communities do provide a window into early evolution. But there are so many uncertainties regarding the conditions for life prior to the advent of oxygenic photosynthesis ($> 3.5 \times 10^9$ years ago) that this approach has limited scope. Furthermore, at least some contemporary anaerobes (pro- as well as eukaryotes) seem to be descended from aerobic ancestors and to have secondarily adapted to an anaerobic existence. Finally, the extant anaerobic biota are maintained and fuelled by the reduced carbon deriving from oxygenic photosynthesis, and they also depend on other interactions with the surrounding oxic world. Many aspects of contemporary anaerobic communities cannot be understood in isolation from the surrounding oxic biosphere and they are therefore likely to differ in fundamental ways from those of the pre-oxic early Archaean.

The book is structured into five chapters. The first serves as a general introduction. It discusses the basic reasons for the presence and maintenance of anoxic habitats in an otherwise oxic biosphere. It also introduces chemical equilibrium considerations and discusses their limitations. The chapter is concluded with a discussion of anaerobic habitats through geological time and a short survey of contemporary anoxic habitats.

The following three chapters constitute the core of the book. Chapter 2 concentrates on different types of anaerobic energy metabolism in prokaryotes (fermentation, anaerobic respiration, anoxygenic photosynthesis and methanogenesis) and the mechanisms which explain coexistence, emphasizing syntrophic interactions. The early evolution of prokaryote energy metabolism is also considered and the conventional wisdom that the fermentation of carbohydrates represents the earliest type of energy metabolism is questioned. The following chapter deals with anaerobic eukaryotes. The current interpretation of their early phylogeny suggests that a few groups are primarily amitochondriate. Most extant anaerobes, however, are descended from aerobes which have secondarily adapted to anaerobic life. Their energy metabolism, syntrophic interactions with symbiotic bacteria, and oxygen toxicity, are discussed in some detail. The chapter is concluded with a discussion of anaerobiosis in metazoa; some aquatic or parasitic species spend frequent or even extended periods when they are fuelled only by anaerobic metabolism, but the existence of metazoa which are entirely independent of access to oxygen for their complete life cycle, has not been documented.

Chapter 4 discusses the structure of different anaerobic habitats and communities in more detail. Emphasis is placed on the spatial and temporal structure, and heterogeneity in time and space. Two sections are also devoted to anaerobic protozoan biota and to the role of phagotrophy in anaerobic communities, respectively.

The last chapter focuses on relationships with the oxic world. The first section deals with the biology of the oxic–anoxic boundary and its significance for the anaerobic biota, stressing the interactions between aerobic and anaerobic communities. The qualitative significance of certain anaerobic processes for biogeochemical element cycling in the biosphere is also discussed. Finally, the ecological and evolutionary significance of oxygen and of oxidative phosphorylation is considered. The arrival of free oxygen also permitted some biosynthetic processes which are characteristic of animals, plants and fungi. Furthermore, the substantial improvement in bioenergetic efficiency allowed for the evolution of longer food chains. While microbial aerobes can manage with a very low partial pressure of oxygen, the evolution of macroscopic animals depended on a high p_{O_2}, approaching that of the present-day atmosphere, and this is related to the apparently sudden occurrence of metazoa in the late Precambrian.

In this book we have taken a broad approach and this decision carries with it some strengths and weaknesses. It was unavoidable that we venture outside our own expertise in several places, but we have tried to compensate for this by providing sufficient references to the original literature. On the other hand, the broad approach will, we hope, appeal to a wider readership and present new perspectives to specialists within any of the topics covered in the book.

1

Anaerobic environments

1.1 Causes and consequences of anaerobiosis

1.1.1 *Operational definition of anaerobiosis*

It is rather difficult to produce a comprehensive and useful definition of 'anoxia'. Part of the problem lies in the different perspectives held within different scientific disciplines: a geologist, an animal physiologist and a microbiologist, for example, will invariably have differing ideas on what constitutes significant oxygen-depletion. Even amongst biologists, there is no concensus on where the 'anoxia' threshold lies. Some of the reasons for this confusion are now largely understood, as will become clear during the course of the book. However, as this is a book about the role of anoxia in the biosphere, it is perhaps useful to have a brief overview of what the term usually means in various areas of science. We shall also define what we will be talking about when we refer to 'anoxia', and 'microaerobic' and 'anaerobic' organisms. The various terms used in quantifying O_2-pressures and solubilities are described in Box 1.1

The definition of anaerobic (anoxic) conditions is complicated by the fact that some microorganisms respire or are otherwise affected at oxygen tensions below the detection limit of conventional quantification methods (Winkler titration and oxygen electrodes have detection limits of 0.5–1% atm.sat. $\approx 10^{-6}$ M). To zoologists and palaeontologists concerned mainly with large multicellular animals the definition of anaerobiosis is much simpler: the respiratory rate of larger animals decreases at p_{O_2}-values around 10% atm.sat. or higher and they convert to a fermentative metabolism (or die) at values well above the detectable level. Geologists consider environments to be 'aerobic' if the oxygen content exceeds > 1 ml ℓ^{-1} ($\approx 18\%$ atm.sat.), to be 'dysaerobic' or 'dysoxic' at concentrations between 0.1 and 1 ml ℓ^{-1} and to be 'azoic' at even lower concentrations. Animal physiologists talk about 'normoxic' and 'hypoxic' conditions; these terms refer to a particular organism, and hypoxic means an oxygen tension which limits the

Box 1.1   Oxygen pressures and solubilities

Under normal atmospheric conditions at sea level, the height of a column of mercury in a mercury barometer is approximately 760 mm. This figure has long been accepted as the 'standard atmosphere' (atm.). The force per unit area exerted by 760 mm of mercury (1 atm.) is 1.01325×10^6 dynes cm^{-2}. One bar is equivalent to 10^6 dyne cm^{-2}, so 1 bar = 0.987 atm. (and hence, in meteorological terminology, 1 atm. = 1013.25 millibar). The SI unit of pressure is the pascal (symbol Pa), defined as a pressure of 1 newton m^{-2} (1 bar = 10^5 Pa).

The Earth's atmosphere contains 20.94% O_2 by volume. Animal physiologists still often quantify partial pressures as mmHg, so the p_{O_2} in the atmosphere is 159 mmHg. In SI units, the equivalent pressure is 21.2 kPa (1 atm. = 1.013 25 bar = 101 325 Pa [× 20.94%]).

The solubility of O_2 in water is low and varies according to temperature and salinity. At atmospheric pressure, the solubility in freshwater at 0 °C and 20 °C is 456 and 291 μM, respectively; in seawater (31 ppt) it is 241 μM at 20 °C. In this book we express p_{O_2} as % atm. sat. (100% = 21.2 kPa) and O_2-concentration in terms of moles ℓ^{-1}.

rate of respiration (below some value between 5 and 50% atm.sat.). 'Suboxic' is used, mainly by geologists, to describe layers in marine sediments which are essentially free of oxygen, but which are not chemically reducing.

Microbiologists have previously used the term 'Pasteur point' to designate the oxygen tension at which facultative anaerobes convert from aerobic respiration to a fermentative metabolism and its value was considered to be 1% atm.sat. However, this is not a universal constant, for it varies considerably between species. Also, many anaerobes are inhibited at much lower oxygen tensions (< 0.1% atm.sat.) while other anaerobes (viz. organisms with an oxygen independent energy metabolism) tolerate even normal atmospheric oxygen tension.

It is also not easy to reproduce strictly anaerobic conditions in the laboratory. Commercial N_2-gas contains traces of oxygen which must be removed chemically (e.g. by passing through a heated copper pipe or mixed with H_2-gas and passed over a catalyst) before it is used for stripping culture media of oxygen or to provide an atmosphere in which organisms can be manipulated in open containers. Much early work on the tolerance to anoxia or attempts to isolate anaerobes have subsequently been shown to be uncritical or unsuccessful due to the presence of traces of oxygen (Brand 1946). The success, during the last two decades, of isolating many novel types of strictly anaerobic bacteria is largely attributable to the development of new and more critical anaerobic techniques such as those of Hungate (1969).

One simple way to remove traces of oxygen is to add strong reducing agents (e.g. dithionite) to cultures or to harness contaminating aerobic bacteria for the removal of the last traces of oxygen. This approach, which mimics what happens in nature, makes it possible to obtain values of p_{O_2} which are far below the limit of detection by quantitative chemical methods. The presence of certain redox-dyes (e.g. resazurin) in culture media makes it possible to detect extremely slight contamination with oxygen.

The classical method of quantifying dissolved oxygen (Winkler titration) is useful for water samples larger than some minimum volume. Many natural anaerobic habitats are so small (e.g. the interior of detrital particles) or characterized by such steep chemical gradients that this method is useless. During the last decades, oxygen microelectrodes have been developed which are mounted in capillary glass tubes or syringes and have an electrode tip measuring down to 5 μm in diameter. These electrodes can map the distribution of oxygen with a spatial resolution of < 0.05 mm (a limit basically set by diffusion). These (and similar microelectrodes for measuring S^{2-}-ion activity and pH) have played a special role in describing anaerobic and microaerobic habitats (Revsbech and Jørgensen 1986; see also Fig. 1.1). Still, the detection limit is not sufficiently low to measure trace-concentrations of oxygen which may affect some microbes.

Two methods, mass spectrometry (Lloyd *et al.* 1981) and the quantification of oxygen-dependent bacterial luminescence (Lloyd *et al.* 1982) have been employed to quantify the lowest detectable p_{O_2}-levels. Such methods have made it possible to determine more accurately the levels at which oxygen uptake by aerobic microorganisms is limited and to demonstrate that the strongly reducing contents of the cow rumen (which has long been a classical example of an anaerobic environment) may periodically have p_{O_2}-values within the range 0.1–1% atm.sat. (Hillman *et al.* 1985).

The conclusion of all this is that the term 'anaerobic' is somewhat relative according to the perspective of the investigator. To microbiologists, 'anaerobic' conditions mean a p_{O_2} which is considerably lower (< 0.1% atm.sat. or 0.2 μM) than the limit of detection by ordinary quantitative chemical or electrochemical methods. The level which affects some strict anaerobic organisms or which totally inhibits biological utilization of oxygen has rarely been determined accurately due to technical difficulties. It must be assumed that in strongly reducing environments (such as sulphide-containing sediments) p_{O_2} is generally extremely low and in practice undetectable. In this book, oxygen tensions within the range 0.1–5% atm.sat. will be referred to as 'microaerobic' (although to a zoologist these levels are for all practical purposes anaerobic).

An organism is considered to be an anaerobe if it has an energy metabolism which is independent of oxygen and if it can complete its entire life cycle in the absence of oxygen. Many aquatic metazoa and protozoa are capable of sustaining themselves for extended periods of time on the basis of fermentation, but they are incapable of completing their life cycle in the absolute absence of oxygen; such organisms are not considered as anaerobes. An organism is a facultative anaerobe if it is capable of oxidative phosphorylation (but access to oxygen is not an absolute requirement for its growth and multiplication). Anaerobes which are incapable of oxidative phosphorylation are referred to as obligate anaerobes. Sensitivity to the presence of oxygen is often considered to be a characteristic feature of anaerobes. While most anaerobes are inhibited above some minimum value of p_{O_2}, this property is, as already mentioned, highly variable. Some are inhibited at levels far below the detection limit of oxygen (e.g. methanogenic bacteria, some sulphate-reducing bacteria); these are often referred to as strict anaerobes.

Sensitivity to oxygen tensions above some threshold value is also a property found in many obligate aerobic organisms both among the prokaryotes and among the eukaryotes. Such organisms are referred to as microaerophiles. Oxygen toxicity is discussed in more detail in Section 3.3.

1.1.2 *Causes of anaerobiosis*

Water emerging at the surface of the Earth by geothermal activity (hot springs, volcanic activity along oceanic ridges) is anoxic and contains reducing solutes such as sulphide. The role of these phenomena in terms of creating anaerobic habitats in the contemporary biosphere is quantitatively trivial; the accumulation of dead organic material is of vastly greater importance. Through oxygenic photosynthesis, plants, algae and cyanobacteria generate local, chemically reducing conditions (in the form of cell carbon) while at the same time oxidizing the environment (through the excretion of oxygen). When the organic material they produce is broken down by aerobic heterotrophs, the redox balance is restored. However, rapid reaction rates in combination with diffusion limited transport of dissolved oxygen may result in local depletion of oxygen. As anaerobic microbial degradation takes over, the metabolic end-products (e.g. sulphide, hydrogen) will render the surroundings chemically reducing. Such events are likely in places where there is a large import of dead organic material and protection from turbulent or advective transport of water. In the contemporary biosphere, anaerobic conditions are primarily due to the local reducing power generated by photosynthesis in conjunction with spatial heterogeneity and diffusion-limited transport of solutes.

An important example is provided by marine and lacustrine sediments. These receive a rain of dead plankton organisms, faecal pellets, etc., which sink to the bottom. In shallow waters, debris of macroalgae and seagrasses are also important. This material is mineralized in the sediment. Here (ignoring burrowing animals or their ventilatory water currents) transport of dissolved oxygen takes place only through molecular diffusion. The consumption of oxygen by animals and bacteria leads to a diffusional flux from the water into the sediment, where it is consumed. The considerations in Box 1.2 (see also Fig. 1.1) predict how the oxygen gradient should look

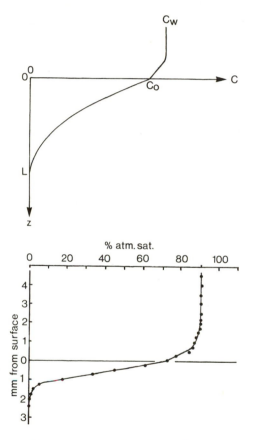

Fig. 1.1 *Above*: the predicted oxygen gradient in a sediment according to eq. [1.1]; the linear decrease in oxygen concentration in the diffusive layer immediately above the sediment is also indicated (C_w: bulk concentration in water, C_0: concentration at sediment surface, L: maximum depth of oxygen penetration). *Below*: the oxygen tension measured at 0.25 mm intervals with an oxygen microelectrode, from 4 mm above to 2.5 mm below the surface of a sediment core (kept in the dark) from Nivå Bay.

Box 1.2   A model of oxygen gradients as a function of respiration and diffusion

At any depth in the sediment (z) the change in O_2-concentration (C) is given by $dC/dt = D(d^2C/dz^2) - R$, where R is the respiration rate per unit sediment volume and D is the diffusion coefficient. At steady state we have $d^2C/dz^2 = R/D$. Integrating, we find the concentration as a function of depth to be given by $C(z) = z^2R/2D + az + b$. The boundary conditions ($C = C(0)$ at the surface ($z = 0$) and $dC/dz = 0$ for $C = 0$) show that the constants $b = C(0)$ and $a = -[2C(0)R/D]^{1/2}$ so that,

$$C(z) = z^2 R/2D - z[2C(0)R/D]^{1/2} + C(0). \qquad [1.1]$$

The concentration gradient will have a parabolic shape. Setting $C = 0$ in [1.1] shows that the depth at which O_2 disappears, $L = z(C = 0) = [2C(0)D/R]^{1/2}$. Since the flux of O_2 per unit sediment surface area, $J = RL\Phi$ (where Φ is sediment porosity), then

$$L = 2D\Phi\, C(0)/J. \qquad [1.2]$$

Knowing the depth at which O_2 disappears and the diffusion coefficient for O_2 in sediments (usually estimated to be about 8×10^{-6} cm^{2}s^{-1}) it is possible to estimate the O_2-uptake of the sediment.

Immediately above the surface of the sediment there is a diffusive boundary layer which is unaffected by the turbulence of the overlying water. Assuming a negligible O_2-uptake in this layer, a linear diffusion gradient will be found. Its slope, $dC/dx = [C(W) - C(0)]/[\text{thickness of diffusive layer}] = J/D$, is also a measurement of the O_2-flux (in water, $D \approx 2.2 \times 10^{-5}$ cm^{2}s^{-1}).

and how the depth of oxygen depletion is determined by the diffusion coefficient of dissolved oxygen and the rate of oxygen consumption.

Figure 1.1 shows that the qualitative prediction of Box 1.1 seems to hold for real sediments, and experience also shows that eq. [1.2] gives a fairly good prediction of directly measured oxygen fluxes. Even so, the implicit assumptions on which these considerations are based are over-simplified. Firstly, it is assumed that no photosynthetic oxygen production takes place in the surface layer; this may lead to a different shape of the oxygen gradient and to deeper penetration of oxygen (Fig. 1.9). It is further assumed that R is invariant with depth; considering the shallowness of the oxic zone this may not be an unrealistic assumption. However, it is also assumed that R is not inhibited by low values of p_{O_2}. This is likely to be true for p_{O_2}-values above the detection level of the oxygen electrode ($\approx 1\%$ atm.sat.), but at

Table 1.1. Time (T) for transport of solutes by molecular diffusion for different distances (L) calculated as $L^2/(2D)$, where the diffusion coefficient (D) for low molecular weight solutes is assumed to be 2×10^{-5} cm^2 s^{-1}

L:	1 μm	10 μm	100 μm	1 mm	1 cm	10 cm	1 m
T:	0.25 ms	25 ms	2.5 s	4.2 min	6.9 h	29 days	7.9 years

lower values, microbial respiration will be inhibited. While the inaccuracy in determining the maximum depth of oxygen penetration will not significantly affect estimates of the oxygen flux, it is likely that (undetectable) trace levels of oxygen occur at greater depths than predicted from eq.[1.1].

At any rate, these considerations demonstrate and explain why in aquatic sediments (with the exception of porous sediments exposed to heavy surf or to tidal pumping) oxygen often cannot be detected a few millimetres beneath the surface. Similar considerations can be made for spherical objects. Thus a detrital particle or a water-logged soil particle will (as seen from considerations similar to those leading to [1.1]) become anaerobic at the centre if the radius $r > [6C(0)D/R]^{1/2}$. Given realistic assumptions about the rate of microbial oxygen uptake inside such particles, it can be predicted that particles as small as 1–2 mm may maintain an anoxic centre, even when the particles are surrounded by air or by oxygenated water.

These considerations present us with some general properties for all anaerobic systems. Anaerobic habitats exist because of the generation of organic material by photosynthetic organisms together with the temporal and spatial heterogeneity of nature. A prerequisite is the accumulation of dead organic matter which is degraded at a rate which exceeds that of the diffusion-limited oxygen supply. Anoxic habitats are always aquatic since diffusion coefficients are about 10^4 times lower than in air. The minimum size of an anaerobic habitat (submerged in aerobic surroundings) is about a millimetre (thus a termite can maintain an anaerobic microbial community in its hindgut). In the present geological period the largest anaerobic habitats are some deep marine basins with overlying stagnant water.

In this book we will repeatedly refer to the importance of molecular diffusion for transport and spatial structure in microbial communities. Table 1.1 gives a feeling for the scale of time and length for diffusion processes.

1.1.3 *Anaerobic environments and reduction–oxidation potentials*

The application of equilibrium considerations to the chemical environment of anaerobic habitats has an heuristic value and, at the same time, offers the possibility of an empirical description of such environments. If the Earth was lifeless, the chemical environment of its surface would be determined by the reactions of volatile substances (appearing through volcanism) with igneous rocks, producing seawater, the atmosphere and sediments. The system would tend towards a chemical equilibrium, but it would never quite be reached, due to kinetic factors, to photochemical processes in the upper atmosphere, and to the more or less continous addition of new volatile substances.

One result of the presence of life is that the surface of the Earth is to a much greater extent in a state of chemical disequilibrium than would otherwise be the case. The contemporary biosphere can be considered as an oxidizing matrix (the atmosphere and the oceans) which contains chemically reducing islands. These are formed by living and dead organic matter and by anaerobic environments. This state of disequilibrium is, as previously discussed, maintained in a dynamic steady state through photosynthesis in conjunction with spatial heterogeneity and kinetic factors.

A redox system can be described by the half-reaction

$$Ox + e^- \rightleftharpoons Red \qquad [1.3]$$

with the concentration reaction quotient $Q = [Red]/[e^-][Ox]$.

Since free electrons do not occur in solution, the reaction [1.3] cannot take place unless it is coupled to another redox system, so that

$$Ox_1 + Red_2 \rightleftharpoons Ox_2 + Red_1. \qquad [1.4]$$

If two half-cells (such as [1.3]) are connected with a 'salt bridge' (e.g. a tube containing a solution of KCl) and platinum electrodes are immersed in each of the half-cells, a voltage difference E, can be measured between the electrodes. (The inert Pt-electrode can be considered as a kind of electron-reservoir and its potential in each half-cell thus depends on the ratio [Red]/[Ox].) If one of the half-cells (the reference-cell) is a hydrogen electrode (pH = 0, 1 atm. H_2, $Q = 1$), then the measured voltage is referred to as E_h of the other cell. If the circuit is closed, a current will pass and the free energy of the system decreases. This loss is identical to the electrical work done by the system, which is $\Delta G = -nFE_h$ per mole (where n is the number of electrons involved in the reaction and F is Faraday's constant). Since $\Delta G = \Delta G^0 + RT\ln Q$ (where ΔG^0 is the change in free energy when transforming 1 mole under standard conditions, R is the gas constant and T absolute temperature) and since in the reference-cell the situation [Red] = [Ox] is maintained, it can be seen that

$$E_h = E_0 + (RT/nF)\ln([Ox]/[Red])$$

where E_0 ($= -\Delta G^0/nF$) is the potential of the half-oxidized system ([Red] = [Ox]). E_0 is a characteristic constant of the system.

In practice, redox potentials are measured with a Pt-electrode against a half cell (such as an Ag/AgCl electrode) using a potentiometer with a high internal resistance ($> 10^{12}\ \Omega$) and the measured potentials are converted to

E_h by adding the potential of the reference electrode (relative to the hydrogen electrode).

Returning to the overall processes of the biosphere we observe that the function of heterotrophic organisms is to catalyse the restoration of chemical equilibrium through the oxidation of reduced carbon produced by photosynthetic organisms (Fig. 1.2). To this end a limited number of electron acceptors are available. From Fig. 1.2 it is apparent that the coupling of the different half-reactions yields different amounts of free energy. For example, the energy yield of oxidizing 1 mole of carbohydrate with oxygen is about six times higher than when sulphate is the oxidant, whereas the oxidation

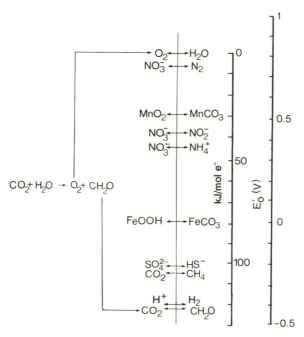

Fig. 1.2 Oxygenic photosynthesis creates a chemical disequilibrium while (respiratory) heterotrophs and methanogens catalyse the return to equilibrium. To this end, a number of external electron acceptors are available, but the gain in energy differs according to the different available redox-couples. Some of the processes also occur spontaneously and do not require biological catalysis (e.g. S^{2-} oxidation by O_2). Phototrophs using electron donors other than H_2O (e.g. S^{2-} or H_2) also create a chemical disequilibrium, but with a lower energetic potential. Fermentation does not in principle imply external electron acceptors (although some processes involving H^+ and CO_2 reductions are included), but it depends on the dismutation of the substrate for energy generation. Since some of the substrates of fermenters do not form reversible redox-couples at a Pt-electrode while the metabolic endproducts often do, fermenters also tend to lower the redox potential (as measured with a Pt-electrode) of the environment.

of methane with sulphate yields only a small amount of free energy. The figure, of course, also shows which processes are thermodynamically impossible (e.g. the oxidation of ferrous iron by sulphate).

Thermodynamical considerations suggest that the energetically most favourable processes should proceed first. Thus, organic material (CH_2O) should first be degraded by oxidation with O_2. Only after O_2 is depleted will NO_3^- serve as an electron acceptor, followed by Mn^{4+}, Fe^{3+}, SO_4^{2-} and CO_2. Biological considerations also suggest that this is so. Living cells which catalyse these processes use part of the free energy they gain for growth and for cell division. The cell yield per unit of assimilated substrate is proportional to the yield of the energy metabolism. Thus, as long as O_2 is present, aerobic respirers will outcompete denitrifying bacteria, etc.

This successional sequence, which is entirely based on equilibrium considerations, is in many respects a realistic description of nature. Deviations occur due to kinetic factors. Also, some potential processes are apparently not realized in any organisms, or they do not proceed spontaneously under ordinary conditions. For example, sulphate reducers cannot utilize carbohydrates directly as substrate, so they depend on fermenting bacteria to convert these compounds into usable substrates (volatile fatty acids or H_2). Some processes may proceed spontaneously (e.g. the reduction of ferric iron with sulphide) and it has not been demonstrated that a biologically mediated Fe^{3+}-reduction in a respiratory process takes place. The different bacterial processes are described in detail in Sections 2.1 and 2.2.

The sequence describes the successional events following, for example, an accumulation of dead organic material: the different environmental electron acceptors are depleted one after the other. There is also a spatial dimension to these considerations. In sediments, oxygen is depleted close to the surface. Below this is a zone characterized by the biologically mediated reduction of nitrate. Manganese and iron also become reduced in this zone by bacterial metabolites. Below this 'suboxic zone', sulphate reduction dominates (the 'sulphidic zone') and beneath this, sulphate is depleted and methanogenesis predominates. The zonation (and deviations from this idealized picture) is described in more detail in Section 4.1.

Measurements of E_h in natural environments and in cultures are of empirical value (yielding information on the predominant processes), but in practice, redox potentials are often difficult or impossible to interpret chemically. This is because some redox couples do not react (kinetic factors prevent equilibrium) so that more than one redox-system is measured simultaneously. Nevertheless, measurements of E_h have been reported frequently in the literature since they represent a simple way of characterizing processes in sediments and other anaerobic systems. Among the first to apply and interpret redox potentials in aquatic environments (lake water and

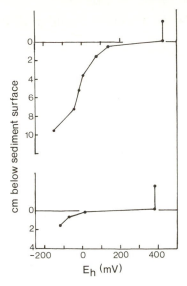

Fig. 1.3 The redox potential in a sandy sediment at 25 m depth in the Øresund (*above*) and in a shallow sediment from Nivå Bay (*below*). In the former case there is a deep 'suboxic' and weakly reducing zone and the sulphidic zone starts at a depth of about 8 cm. In the latter case, free sulphide is present almost at the sediment surface (data from Fenchel 1969).

sediments) were Hutchinson *et al.* (1939), Pearsall and Mortimer (1939) and Mortimer (1941–42); for a review of the earlier literature see also Fenchel (1969).

A Pt-electrode does not form a reversible oxygen electrode when immersed in water. Electrode potentials measured in oxygenated seawater are within the range 400–470 mV (rather than around 800 mV as would be expected) and these values are almost unchanged if the water is stripped of oxygen by flushing with N_2. If the electrode is inserted into sediment (Fig. 1.3) the potential rapidly falls to values within the range of 0–200 mV. These are generally believed to be buffered by the Fe^{3+}/Fe^{2+} system. When penetrating deeper, values fall to around −200 mV; this is believed to be controlled by the system $S^{2-} \rightleftharpoons S^0 + e^-$ and there is a close correlation between E_h and S^{2-}-ion activity (Berner 1963). Lower values are not found in sulphidic systems. In sediments with a high load of organic matter, the lower part of the oxygenated zone overlaps with the upper part of the sulphidic zone (low concentrations of HS^- coexist with low concentrations of O_2). In this zone E_h falls steeply to around −200 mV so only one distinct 'redox discontinuity layer' (Fenchel 1969) is present. Lower values of E_h (around −300 mV) can be recorded in non-sulphidic systems (e.g. the cow rumen) and this can probably be attributed to H_2.

1.2 Anaerobic habitats through geological time

1.2.1 *The earliest atmosphere and the origin of life*

From an historical point of view the earliest contributions to our understanding of the chemical environment of the early atmosphere derive from speculations on how, and under which conditions life arose on Earth. In 1924, Oparin (Oparin 1953) and, independently, Haldane (1928) argued that the first organisms were fermenters which utilized abiotically produced organic compounds. Fermentation seemed to be the simplest (and at the time the only fully understood) type of energy metabolism. In addition, organic molecules would be stable only in the absence of molecular oxygen. By implication, the presence of oxygen in the atmosphere was secondary and due to the evolution of oxygenic photosynthesis at some later time.

These ideas received further attention and support through Miller's (1953) demonstration that in a hypothetical, O_2-free primordial atmosphere (assumed to include H_2O, CH_4 and NH_3) exposed to electrical discharges (mimicking Precambrian lightning) a variety of organic molecules and their polymerization products are obtained. The most striking product was a high yield of amino acids, principally those which are the 'essential' amino acids constituting the proteins of extant organisms (among a very high number of chemically possible amino acids). Experiments on 'prebiotic chemistry' have since evolved into a discipline of their own. Extended experimentation has shown that different plausible compositions of a primordial atmosphere plus an energy source (such as electrical discharges or short wavelength light) led to the synthesis of a wide variety of organic molecules including purine and pyrimidine bases, nucleotides, porphyrins, carbohydrates and low-molecular weight fatty acids (for reviews see Chang *et al.* 1983; Miller and Orgel 1974; Schuster 1981). This work has led to a popular view of the early Earth as being covered with a 'primordial soup'; a broth consisting of a variety of biologically essential low-molecular weight organic compounds and polymerization products (a situation which may today be realized in the atmosphere of Jupiter and on the surface of Saturn's moon Titan). Although many aspects are unclear, including the actual composition of the primordial atmosphere (see below) and the concentration of organic molecules which could actually have accumulated, the results of this work provide a convincing picture of the chemical environment in which life originated.

It is, however, necessary to point out that the abiotic synthesis of the essential molecules of life does not represent an explanation of the origin of life itself. Some authors have focused on various energy-yielding processes and the evolution of energy metabolism and used this as a cornerstone for theories of the origin and evolution of life, and they have assumed that the genetic system is secondary. To us this does not seem convincing. As far as the origin of life is concerned, the essential problem is the explanation of the origin of a self-replicating system which is subject to Darwinian selection, and the origin of the phenotype–genotype duality. The most fruitful model seems to be the evolution of self-replicating RNA-molecules ('quasi-species' resembling tRNA-molecules) and their coupling in auto-catalytic cycles ('hypercycles') compartmentalized inside lipid membranes. This theory is attractive for many reasons and not least because such systems of *in vitro* replicating short RNA-molecules have been subject to experimentation with some success (Cech 1985; Eigen 1992; Inoue and Orgel 1983; Orgel 1986; Schuster 1981).

How evolution took place from an 'RNA-world' and to the first cell (with DNA as the carrier of genetic information and with more or less complex biochemical pathways involved in energy and assimilatory metabolism) is still a matter of speculation which is only mildly constrained by facts. Molecular evidence from extant organisms is not inconsistent with the assumption that the first cells (or at least the common ancestor of eubacteria, archaebacteria and eukaryotes) lived about 4 billion ($= 4 \times 10^9$) years ago. The oldest sedimentary rocks are about 3.8 billion years old and leave no evidence of the type of life then present. Regarding metabolism, it must be assumed that the first self-replicating molecules depended on energy from purine and pyrimidine bases with energy-rich phosphate bonds which were synthesized photochemically in the environment, and that all other building blocks were also available in the 'primordial soup'. It is still widely believed that a *Clostridium*-type fermentation of carbohydrates represents the most original type of energy metabolism (e.g. Gest and Schopf 1983). However, it seems as likely that the first type of energy metabolism (as environmental energy resources became scarce) was represented by photophosphorylation and a membrane-bound ATPase (e.g. Broda 1975; Pierson and Olson 1989). Certainly, phylogenetic trees based on rRNA-sequences of extant prokaryotes show that phototrophic metabolism (and chlorophylls) must have an ancient property which was shared by the ancestors of most eubacterial groups (Woese 1987 and Section 2.4).

There does seem to be a general consensus that:

(1) phototrophy was a very early development;

(2) respiration developed from phototrophy (supported by, for example, the common [non-heme Fe/S enzymes – quinone – cytochrome b] sequence of electron-carriers in photophosphorylation and respiration);

(3) several types of fermentative energy metabolism evolved from respiring or phototrophic forms;

(4) anoxygenic photosynthesis (e.g. using sulphide, organics, ferrous iron or hydrogen as electron donors) predated oxygenic photosynthesis; and

(5) anaerobic respiration (e.g. using sulphate as electron acceptor) predated aerobic respiration.

Both anaerobic and aerobic respiration seem to have developed independently in several groups of bacteria. Methanogenic bacteria are generally believed to represent an ancient type of energy metabolism; their existence, however, depends on a source of H_2 (or acetate). Whether appreciable amounts of these substrates were available prior to their production by fermenting bacteria remains an open question. The evolution of prokaryote energy metabolism will be considered in more detail in Section 2.4.

Beyond these considerations, the chain of evolutionary events and the quantitative role of various energy-yielding processes prior to the advent of oxygenic photosynthesis are quite open to speculation. This is because the oldest traces of life and other geological evidence (see below) dating from about 3.5 billions (3.5×10^9) years ago suggest that oxygenic photosynthesis had developed by then. Also, as explained in the following section, geophysical considerations still allow for a wide range of physical and chemical conditions during the first 0.8 billion years of the history of the Earth.

1.2.2 *Geophysical considerations on the composition of the early atmosphere*

There is strong geological evidence to show that during the early Precambrian the p_{O_2} of the atmosphere was extremely low (see the following section). There is also general agreement that the source of the primordial atmosphere (outgassing from the earth or from accreting meteorites) could not have contained oxygen and that the small amounts of oxygen produced photochemically in the atmosphere by photolysis of water would be maintained at a very low level due to weathering processes. It is, however, debatable just how reducing (viz. the magnitude of the H : O ratio) the atmosphere was. Equilibrium as well as kinetic considerations are necessary in order to arrive at any conclusion regarding the evolution of the atmosphere. Unfortunately, the boundary conditions for such considerations are still so poorly known that they allow for a wide range of possibilities.

Holland (1984) gave a comprehensive discussion of the problem; other recent accounts are given by Kasting (1993), contributions in Schopf (1983), and in Schopf and Klein (1992). The primordial atmosphere is believed to have originated through outgassing from the Earth. Today, volcanic gases consist mainly of N_2, H_2O, SO_2, and CO_2 with only small amounts of H_2 and would thus be responsible for a weakly reducing atmosphere. However, two possible scenarios could have resulted in a more reducing atmosphere. If a larger part of the iron content of the Earth would then still have been present in the mantle (and the core formed later) the outgassing would mainly (in addition to N_2 and H_2O) have contained H_2, CO and H_2S. Alternatively, a relatively rapid (10^7–10^8 years) accretion of the Earth would have resulted in a hot, molten surface and a release from the accreting material during infall; this could also account for a more strongly reducing primordial atmosphere. An anoxic, but only weakly reducing atmosphere could be the source of organic material, via the synthesis of formaldehyde and cyanide; however, the yield would be much lower than that of a more reducing 'Miller-atmosphere'. The presence of methane and ammonia in the atmosphere would result from photochemical processes between H_2 and CO and N_2, respectively. In the absence of any hard evidence regarding the redox-conditions of the original atmosphere, the experiments on prebiotic chemistry in conjunction with the fact that life actually did appear, is an argument in favour of the view that the earliest atmosphere was reducing.

Irrespective of the exact initial conditions of the atmosphere it tended to oxidize slowly over geological time due to the loss of H_2 (primordial or deriving from photolysis of H_2O) to space. If life had not evolved on Earth, the present atmosphere would probably be like that of Mars: oxidizing, but anoxic since all free O_2 produced would rapidly be bound, mainly to Fe, through weathering processes. There are good reasons to assume that the p_{CO_2} was much higher in the atmosphere of the Earth during the early Precambrian and that it has slowly, but irregularly decreased to the present level. It is not known how much higher the value was. One argument in favour of a high CO_2-concentration is that the luminosity of the early Sun was probably about 20% lower than it is today and a greenhouse effect would account for the apparent absence of ice cover during the early Precambrian.

Any opinion on the important biogeochemical processes and biogeochemical cycling prior to the advent of oxygenic photosynthesis (before 3.5×10^9 years ago) can only be speculative. It is conceivable that after the (probably short) period of abiotic synthesis of organic molecules, biological processes must have been driven by anoxygenic photosynthesis. As pointed out by Walker (1980) it sounds paradoxical, but anaerobic life may have had a hard time prior to the advent of oxygenic photosynthesis due to the scarcity of available reducing power to sustain anoxygenic photosynthesis. Hydrogen and low-molecular weight organic molecules would have been scarce. Sulphide (sustaining photosynthetic sulphur bacteria) would seem a possible important source, but according to Walker (1980) this would have been bound largely as metallic sulphides and not necessarily freely available. Also, sulphate-containing evaporites are rare in early Precambrian deposits and the earliest indirect evidence for the activity of photosynthetic sulphur bacteria (in the form of barite deposits) is contemporary with the earliest evidence of oxygenic photosynthesis. The earliest direct evidence for sulphate respiration (in the form of the stable sulphur isotope ratio of iron sulphide deposits) comes from 2.7×10^9 years ago (Gest and Schopf 1983). Even so, it is probable that a complete biogeochemical sulphur cycle including phototrophic sulphur bacteria (utilizing reduced S-compounds for producing organic carbon) and sulphate reducing bacteria (using oxidized S-compounds for respiration) existed prior to oxygenic phototrophic bacteria.

Walker (1980) also suggested that ferrous iron (which had to be an abundant constituent of anaerobic seawater) could serve as reducing power for early phototrophic organisms. This idea is supported by the more recent finding that some contemporary anoxygenic photosynthetic bacteria are actually capable of using ferrous iron as electron donor in photosynthesis (Widdel *et al.* 1993).

Finally, it is possible that during this early stage in the evolution of life, high biological activity was localized mainly in the vicinity of shallow-water sites with volcanic outgassing, and shallow-water hydrothermal vents. Here, the availability of reducing power in the form of sulphide and hydrogen, together with the necessary light intensity could produce centres of high biological activity in seas with an otherwise low biological activity. Conceivably, volcanic activity was more common on the surface of the young Earth since heat production in the mantle, due to radioactive decay, must have been more intense than it is today. But it does not seem unlikely that the use of the ubiquitous electron donor water, by photosynthesizing organisms, made possible the maintenance of a high biomass and a high intensity of biological processes.

1.2.3 *Geological evidence for evolution of the Precambrian atmosphere and the biota*

Two impressive volumes (Schopf 1983; Schopf and Klein 1992) have recently reviewed our current knowledge of the Precambrian and especially the Precambrian biota; the present discussion is based mainly on these works. The most important factual evidence of Precambrian evolution is shown in Fig. 1.4.

The age of the Earth is about 4.6×10^9 years. The oldest known rocks on the Earth are the ≈ 4 billion years old remains of continental crust showing that the surface had solidified, a process which is believed to have happened about 200 million years earlier. The oldest (≈ 3.8 billion years) sedimentary rocks have been found at Isua on the west coast of Greenland showing that continents existed and that a normal hydrological cycle was operating (this also constrains temperatures to between 0 and 100 °C unless atmospheric pressure was much higher than today). It has not been possible

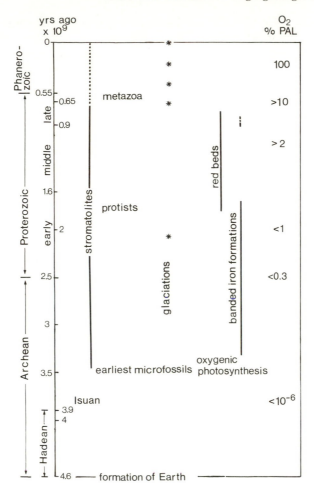

Fig. 1.4 The most important factual evidence regarding the evolution of Precambrian life and the probable atmospheric p_{O_2} (compiled from data in Schopf and Klein 1992).

to determine the provenance or significance (in terms of the existence or properties of contemporary life) of kerogens (organic matter) extracted from this geological formation.

The earliest convincing evidence for life derives from the $\approx$ 3.5-billion-years-old Fig Tree Formation, Onverwacht in southern Africa and from the contemporary Warrawoona formation in western Australia. These consist of laminated dolomitic cherts or 'stromatolites' (see below). Stable isotope analysis of carbon from extracted kerogen suggests that it is photosynthetic in origin. Most importantly, the formations have yielded what are now accepted as *bona fide* microfossils. These include cellular filamentous forms and sheathed colonies. These fossils resemble contemporary blue-green bacteria (e.g. *Phormidium, Rivularia* and *Chlorococcus*; Schopf 1992; Schopf and Walter 1983). Stromatolites are laminated structures found from this period onwards; they are most abundant during the Proterozoic period. They become scarcer towards the end of the Precambrian, but they are also known from the Phanerozoic•geological record. It is now known that the structures are fossilized remains of microbial mats, the matrix of which is made by filamentous, photosynthetic bacteria. Extant analogues are known from many shallow-water habitats which are not exposed to larger burrowing or browsing benthic invertebrates and they are today especially common in hot or hyperhaline environments (Cohen and Rosenberg 1989). The appearance of larger metazoa in the Vendian (about 650 million years ago) explains why they became rarer during this period. (Extant stromatolites are discussed in more detail in Sections 1.3 and 4.1.) The spatial orientation of the microbial filaments of these oldest (3.5 billion years) stromatolites shows (in analogy with properties of extant microbial mats) that the organisms were phototactic. Desiccation cracks and occasional evaporites suggest that these oldest stromatolites formed in shallow waters and were occasionally subject to desiccation (Walter 1983).

Contemporary microbial mats based on filamentous bacteria with anoxygenic photosynthesis (*Chloroflexus*) are known (Ward *et al.* 1984). It is therefore possible that the photosynthetic filamentous organisms of Onverwacht and Warrawoona were also not oxygenic. However, among contemporary prokaryotes, the cell diameter of the oxygenic cyanobacteria is characteristically larger than that of most other forms. Using this criterion, many of these middle Archean forms were cyanobacteria (Schopf 1992). This, together with the fact that the earliest 'banded iron formations' formed ≈ 3.5 billion years ago, constitutes strong evidence that oxygenic photosynthesis took place at that time.

Banded iron formations are huge laminated deposits containing partly oxidized iron (the ratio $Fe^{3+}/(Fe^{2+} + Fe^{3+})$ is between 0.31 and 0.58) in the form of FeO and Fe_2O_3. They were formed in the period between the mid Archaean and the end of the early Proterozoic (3.5–2 billion years ago). Most authors agree that these widespread formations reflect the presence of a low level of oxygen in the atmosphere, but the exact way in which they formed is still debated. The most accepted model assumes that the seas were still basically anaerobic, but that the atmosphere and the surface waters contained small amounts of oxygen. Upwelling would bring anoxic water with dissolved ferrous iron to the surface where it would be oxidized to insoluble ferric iron which was then deposited in shelf seas. The laminated structure has been interpreted as a sign of seasonality. Widdel *et al.* (1993) have suggested that phototrophic oxidation of ferrous iron in the surface waters could have been responsible. This mechanism would not involve the presence of free oxygen; however the explanation is perhaps less convincing, since banded iron formations contain only small amounts of organic material and a scarcity of microfossils (Schopf and Klein 1992). After about 2 billion years ago the banded iron formations were replaced by deposits referred to as red beds. These are alluvial deposits containing ferric iron and they signify increasing atmospheric p_{O_2}.

Regarding the evolution of atmospheric oxygen, certain facts constrain the possibilities. Marine deposits and palaeosols older than about 2 billion years contain the minerals pyrite (FeS_2) and uraninite (UO_2) which show signs of abrasion and weathering. In the present atmosphere this is not possible since these minerals would oxidize and dissolve. Their presence indicates that atmospheric p_{O_2} was < 1% PAL (present atmospheric level), the precise upper limit being somewhat dependent on the (unknown) magnitude of atmospheric p_{CO_2}. Prior to the mid Archean ($\approx$ 3.5 billion years ago) atmospheric p_{O_2} must have been very low. Atmospheric p_{O_2} must have increased towards the end of the early Proterozoic ($\approx$ 2 billion years ago) because of the absence of uraninite in sedimentary rocks and the appearance of the red bed formations. Around 650 million years ago (towards the end of the Vendian) atmospheric p_{O_2} must have reached a level of at least 10% PAL since larger metazoa appear as evidenced especially from the fossils from the Australian Ediacara formation (Glaessner 1984; Nursall 1959; see also Section 5.3). Figure 1.5 shows the possible range of atmospheric p_{O_2} through geological time.

The precise mechanisms by which atmospheric p_{O_2} was (and is) controlled are still not well understood. The oxygen which was produced by the first cyanobacteria must have been consumed by oxygen-respiring organisms (which must have evolved rapidly after the advent of oxygenic photosynthesis) but also by weathering (especially by the oxidation of ferrous iron as is evident from banded iron formations and red beds) and the oxidation of volcanic gases. The subsequent accumulation of O_2 in the atmosphere ($+SO_4^{2-}$ in seawater) must then largely have balanced the accumulation of fossilized organic matter in the form of kerogen (+ pyrite) in sedimentary rocks. The kerogen content of Proterozoic shales is similar to that of Phanerozoic shales, suggesting that primary production was of the same magnitude as now. But the factors limiting continuous increase in atmospheric oxygen are not clear. One suggestion is that reactive phosphate limits primary production and since phosphate is precipitated under aerobic conditions, but released under anaerobic conditions, this might represent a control mechanism. The colonization of land in the Silurian probably introduced new forms of control: it is possible that in the later period of the Phanerozoic, atmospheric p_{O_2} was limited by the frequency of forest fires (for further discussion see Schopf and Klein 1992).

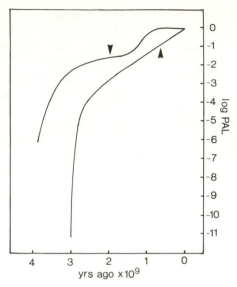

Fig. 1.5 Upper and lower limits to the possible atmospheric p_{O_2} during geological time. The arrows indicate the upper limit of 1–2% PAL around 2 billion years ago, prior to which uraninite and pyrite grains occur, and the lower limit of about 10% PAL during the Vendian (about 600 million years ago) when larger metazoans appear (compiled from data in Schopf and Klein 1992).

One aspect of the evolution of atmospheric oxygen has been somewhat neglected, but is evident from the study of extant microbial mats (stromatolites). While the deeper (> 5 mm) aphotic layers (which have high heterotrophic activity) are anaerobic, the surface layers, when illuminated, build up oxygen tensions which may exceed 300% supersaturation (Section 4.1 and Fig. 1.9). This is likely to have been the case also in the earliest stromatolites, so while the p_{O_2} of the atmosphere was very low and the oceans were still basically anoxic, extremely high oxygen tensions must have been localized in stromatolites since the mid Archaean. Exposure to high p_{O_2} must therefore have been an ecological and evolutionary factor since mid Archaean times.

Stromatolites represent the earliest known biotic community and they have existed continuously for 3.5 billion years. There is no reason to assume that Precambrian microbial mats differed from those of the present day with respect to composition, structure or function, but they had a much wider global distribution. The well-preserved microfossils from Proterozoic times (e.g. the Gunflint formation from North America and similar deposits in Australia) present microbial forms which are apparently identical to extant relatives. Larger fossil cells from around 1.4 billion years old deposits and onwards have been interpreted as the remains of eukaryotes, and in the late

Proterozoic remains of eukaryote phytoplankton have frequently been found (Schopf and Klein 1992). Eukaryotic unicellular organisms are probably under-represented in the fossil record since many or most extant species do not have cell walls or other skeletal material, so they leave no recognizable remains.

Some other geological evidence is relevant in the business of trying to understand Precambrian evolution. For example, the tendency of some biological processes to discriminate between different stable isotopes of some elements has been used to establish the existence of certain processes. Photosynthetic organisms discriminate against $^{13}CO_2$ (relative to $^{12}CO_2$) and the isotope ratio of C in kerogen is used as evidence for a photosynthetic origin. Similarly, the stable isotope ratio of sulphur has been used for evaluating whether sulphide deposits are of biological origin. Progress has been made (discussed in Schopf and Klein 1992), but one persistent difficulty when interpreting the data is lack of knowledge on the 'openness' of the studied systems; the significance of this is evident from isotope distributions in extant systems (e.g. Leventhal 1983). The extraction and identification of different types of biological molecules ('biomarkers'), which are specific for certain taxonomic or functional groups of organisms, faces many difficulties and some of the earlier work is probably suspect because of the prevalence of artefacts and contamination. The study of hydrocarbons deriving from membrane lipids has demonstrated the presence of archaebacteria and eubacteria in Precambrian kerogens and evidence for eukaryotic cell membrane lipids has been found in 1.7 billion-year-old deposits (Schopf and Klein 1992).

The main conclusion based on the available evidence concerning Precambrian environments and biota is that practically all of the basic evolution in terms of biochemistry and types of energy metabolism, had taken place earlier than 3.5 billion years ago, during periods for which we have no direct evidence. On the other hand, a uniformitarian view appears regarding the biosphere since the mid Archaean, where direct evidence is available. The overall concentration of oxygen was, at least initially, very low in the atmosphere and in the oceans (but it could be locally high in cyanobacterial mats). Eukaryotes (at least in a modern form with mitochondria) may have first appeared about 1.5 billion years ago, and larger metazoa and calcareous macroalgae first appeared in the Vendian only about 100 million years before the start of the Cambrian. But all basic types of bioenergetic processes probably existed 3.5 billion years ago and the biogeochemical cycling of carbon, nitrogen and sulphur was established as we know it today, although the relative importance of the different element cycles could have been different. Most striking is the fact that what seems to have been the dominant biotic community (stromatolitic microbial mats) has existed unchanged until this time and can be studied experimentally today. While it is true that the early evolution of life took place on an anaerobic Earth and while most habitats (the oceans) were probably still anaerobic or almost so in the mid Archaean, the reducing power which sustains anaerobic communities (the substrates for anaerobic heterotrophs and for anoxygenic phototrophs) was already then largely supplied from oxygenic photosynthesis.

1.2.4 *Anaerobic habitats during the Phanerozoic era*

At the beginning of the Cambrian period, the p_{O_2} had risen to at least 10% PAL and it could have been even closer to the present value. Even so, during some periods in the last 550 million years, extensive areas of the shelf seas seem to have been anaerobic below some depth, probably not unlike the Black Sea today. The evidence for this is the widespread occurrence of black shales which are rich in organic matter and reduced sulphur. Anaerobic conditions during deposition can also be deduced from the presence of well-preserved fossils such as of the planktonic graptolites (since bioturbation and scavenging by larger animals were absent). These black shales are especially common in older (Cambrian to Devonian) Palaeozoic strata. Similar phenomena are also known from other periods and especially from Jurassic and Cretaceous deposits (Tyson and Pearson 1991). Widespread shelf sea anoxia has also been implied as a contributing factor for the Permo-Triassic extinctions of marine invertebrates (Erwin 1994).

The reason why the deeper layers of shelf seas and possibly the oceans tended to become anoxic in certain periods during the Phanerozoic era has been debated. It is, of course, possible that in the case of the older Palaeozoic deposits a still relatively low atmospheric p_{O_2} played a role, since at saturation seawater would contain less oxygen which therefore became depleted earlier, but this explanation cannot be valid for similar later events. Berry and Wilde (1978) and Wilde and Berry (1986) have suggested that the events resulted from changes in major ocean currents during periods with a warm climate. Today the Earth has a relatively cold climate with ice-covered poles. Warm, oxygenated surface currents move towards the poles, cool and sink and flow back towards the equator, and thus maintain oxygenated deep oceans. In relatively warmer geological periods, this ventilation mechanism is absent. The superficial layers of the oceans remained oxygenated due to photosynthesis and contact with the atmosphere, but the deeper parts would tend to become anoxic.

1.3 Contemporary anaerobic habitats

Anaerobic habitats have existed continuously throughout the history of the Earth. As shown in the previous section, the atmosphere and the seas were initially anoxic, a situation maintained by weathering processes which removed any photochemically produced oxygen. After the advent of oxygenic photosynthesis, the atmosphere, and gradually also the oceans became oxic and the reducing power of organic material became important in maintaining local anaerobic habitats.

In this section, different types of contemporary anaerobic habitats will be examined. Special emphasis is given to the classification of such environments and the basic mechanisms which maintain them.

1.3.1 *The anaerobic water column*

Contemporary oceans are oxic. However, an 'oxygen minimum zone' is found at depths between about 100 and 1000 metres; this is especially pronounced where there is a high rate of production, such as adjacent upwelling areas. The reason for this is the depletion of oxygen due to the mineralization of sinking detrital material below the mixed upper layers of the water column. Below the oxygen minimum zone, p_{O_2} increases again due to a decreased oxygen demand and the deep circulation of oxygen rich water deriving from high latitudes. The phenomenon is more pronounced in the Indian Ocean and especially in the north-eastern Pacific than in the Atlantic Ocean, which receives deep water from both the Antarctic and the Arctic regions. In some areas of the Pacific, oxygen tensions may reach the limit of detection at some depth (Kester 1975).

For anoxic conditions to develop in the water column it must be protected from the import of oxygen containing water by advective transport or eddy diffusion. This may occur in basins delimited by a sill in conjunction with vertical stratification of the water column by a pycnocline (vertical temperature and/or salinity gradients).

The largest contemporary anaerobic water mass is the Black Sea (Deuser 1971, 1975; Emery and Hunt 1974; Grasshoff 1975; Sorokin 1972) which, everywhere below the aerobic surface layer of 100–240 m (according to location), is anoxic and contains sulphide (Fig. 1.6). The current state of the Black Sea seems to have developed about 7000 years ago. Saline Mediterranean water has passed the shallow Bosporus strait since the end of the last glacial period and in the brackish Black Sea, this saline water sinks to the bottom. This causes a vertical stratification of the water column, most of which never comes into contact with the atmosphere. The vertical flux of organic particles produced in the photic zone maintains anoxic and reducing conditions in the deeper waters.

The Cariaco Trench, situated north of Venezuela in the Caribbean Sea, is another large anoxic basin. It was discovered in 1956. Its horizontal dimensions are about 200×50 km and it is about 1400 m deep. A sill prevents horizontal transport of deep water and the water column is

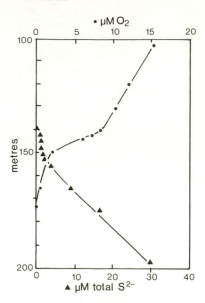

Fig. 1.6 The distribution of O_2 and of total S^{2-} between 100 and 200 m depth in the Black Sea (data from Sorokin 1972).

vertically stratified by a pycnocline; it is anoxic beneath about 380 m depth (Richards 1975; Zhang and Millero 1993). Several such anoxic basins are also known from the Baltic Sea (Grasshoff 1975; Rheinheimer *et al.* 1989). The Baltic is a brackish sea (the larger, central part has a salinity of around 6 ppt). At irregular intervals of several years, special weather conditions result in larger amounts of seawater being transported through the inner Danish waters (the Belt Sea and the Sound) and into the Baltic, where the saline water accumulates in the deeper basins. These vertically stratified basins are therefore anaerobic although the extent of the anaerobic conditions has varied during the last century.

Smaller scale anaerobic marine basins are common throughout the world. Many Norwegian fjords with a sill at the entrance have permanently anaerobic deep water (Grasshoff 1975; Indrebø *et al.* 1979) and similar phenomena are known from relatively shallow basins in 'fjords' of glacial origin such as in the inner Danish waters and along the German Baltic coasts (Fenchel *et al.* 1990; Grasshoff 1975; see also Fig. 1.7). Such localities sometimes resemble temperate lakes in that the thermal stratification breaks down during autumn and winter, so anoxic conditions become a seasonal event (Jørgensen 1980; Kemp *et al.* 1992).

Most lakes, except for very shallow ones, tend to become thermally stratified during a part of the year, and productive lakes consequently develop anoxic deep waters. In temperate climates lakes tend to stratify

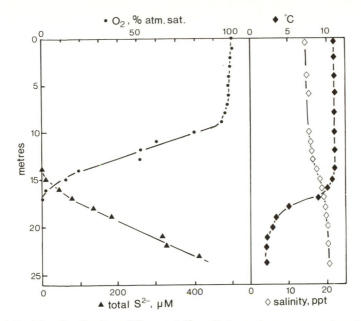

Fig. 1.7 The distribution of O_2, total S^{2-}, salinity and temperature in a 25 m deep depression in the otherwise shallow Mariager Fjord (Denmark), October 1987 (data from Fenchel *et al.* 1990).

during the summer; they may be 'monomictic' (in which case the stratification breaks down during autumn and the water column remains mixed until thermal stratification builds up in spring) or they may be 'dimictic' (in which case an inverse thermal stratification builds up during winter. During stratification, the mixed surface layer is referred to as the 'epilimnion', the layer beneath the thermocline is the 'hypolimnion', while the layer which includes the thermocline is referred to as the 'metalimnion'. The hypolimnion of most biologically productive temperate lakes is anoxic during the summer (Fig. 1.8) and the disappearance of oxygen typically takes place in the metalimnion. At lower latitudes the vertical stratification of lakes is less seasonal; according to depth, area/depth ratio and wind exposure they may be 'polymictic' or 'oligomictic'; in all cases, such stratification also typically leads to an anoxic hypolimnion which may be subject to more or less frequent episodes of mixing and re-oxidation (Wetzel 1975).

'Meromictic' lakes have a permanent vertical stratification which is maintained by a salinity gradient. These more specialized lakes occur when deposits of gypsum or salt are dissolved in the deeper part of the water column. Such lakes often have a permanent anoxic hypolimnion (e.g. Miracle *et al.* 1992; see also Fig. 4.8).

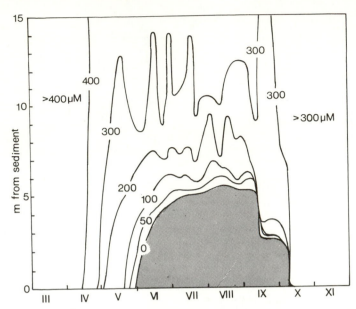

Fig. 1.8 The development of an anoxic hypolimnion following summer stratification followed by mixing of the water column during autumn turnover in Esthwaite Water (English Lake District) in 1980 (data from Finlay 1981).

In principle it is possible to model the vertical distribution of oxygen in a stratified water column on the basis of considerations similar to those leading to [1.1] which describes the vertical penetration of oxygen in a sediment. Such attempts have been made (see e.g. Kester 1975), but in practice this is much more difficult. There are several reasons for this. In the water column molecular diffusion of solutes is unimportant relative to eddy diffusion and vertical advection, and the magnitude of these effects decreases with depth above the pycnocline. Transport of 'parcels' of water across the pycnocline may take place due to instabilities of internal waves which again depend on wind force and direction. Oxygen uptake will occur throughout the upper oxygenated water column, but it will depend on depth and most of the oxygen of the water column is typically consumed through sulphide oxidation at the oxic–anoxic transition layer. This can be seen from Figs 1.6 and 1.7 where the oxygen gradient becomes more 'concave downwards' in this zone whereas it is almost linear in the relatively un-stirred water layer immediately above (in the mixed zone p_{O_2} is almost constantly about 100% atm.sat.). Altogether, the depth dependency of transport rates and the role of lake geometry and weather conditions make it difficult to predict the exact depth of the oxic–anoxic transition zone and the thickness of the zone at which O_2 and HS^- coexist.

1.3.2 Sediments

Sediments constitute the most widespread and continuous anaerobic habitats. With the exception of heavily exposed porous (sandy) beaches ('high energy beaches', see Fenchel and Riedl 1970) aquatic sediments are always anaerobic at some depth. In contrast to the water column, transport of dissolved oxygen is first of all due to molecular diffusion, and the description presented in Section 1.1 of how p_{O_2} decreases with depth is almost complete. However, the depth at which the interstitial water becomes anaerobic depends, in addition to the respiratory oxygen demand (viz. the input of organic matter), also on light, and on bioturbation.

In off-shore localities, at depths with insufficient light for photosynthesis, the anaerobic–aerobic boundary is first of all determined by the input of organic material which sinks to the bottom from the productive photic zone of the water column. In shelf seas (< 200 m), up to 50% of the production ends on the bottom. Typically, the anaerobic/aerobic boundary layer is situated 1–6 mm beneath the sediment surface (Jørgensen 1983; Jørgensen and Revsbech 1989). However, burrows of worms, molluscs and other invertebrates, which may extend many centimetres beneath the surface, are surrounded by narrow, oxic zones (Aller and Yingst 1978; Fig. 4.15). In the oceans, productivity is typically lower and only about 1% of the primary production of the surface waters reaches the bottom at 5000–6000 m depth (the remainder being mineralized in the water column). Consequently, the oxic zone of deep-sea sediments is much thicker (up to 50 cm; see Jørgensen 1983). It is characteristic for sediments with a relatively low input of organic matter that the suboxic zone (in which oxidized Fe, Mn and nitrate are the main electron acceptors) situated between the oxic and the sulphidic zones is also relatively thick.

Beneath areas of high productivity, the anaerobic and sulphidic zone may extend almost to the surface of shelf-sea sediments. This becomes apparent as a white covering layer of sulphide oxidizing bacteria (notably species of *Beggiatoa* and *Thioploca*); see Section 5.1 and Figs 5.1 and 5.2. Such sediments covered by white sulphur bacteria are widespread at 50–250 m depth along the upwelling zone off the coast of Peru and Chile (Gallardo 1977). A similar phenomenon is sometimes seen in the inner Danish waters following the planktonic diatom bloom in early spring (unpublished).

Shallow-water sediments differ in that they typically import more organic material, not only in the form of detrital material from the plankton, but also as debris derived from seagrasses and macroalgae and not least from primary production in the surface layers due to cyanobacteria, diatoms and other eukaryotic algae. In sediments, photosynthesis extends down to a little more than 3 mm depth (Fenchel and Straarup 1971; Revsbech *et al.* 1980). Oxygenic photosynthesis provides substrates for aerobic and anaerobic heterotrophs. At the same time, oxygen is produced, so when the sediment is illuminated the aerobic zone extends downward, to about 4 mm, and strata which are anaerobic in the dark may become supersaturated with oxygen in the light. The oxidized zone therefore shows a diurnal vertical migration which is reflected in vertical migrations of the microbial biota (Section 4.1).

Protected productive shallow-water sediments are often anaerobic almost to the surface; this shows as white patches of chemolithotrophic sulphur bacteria or purple patches of phototrophic sulphur bacteria (Fig. 5.1). During calm nights the anaerobic zone may extend into the overlying water column. Conversely, in wave-exposed, sandy sediments some advective water flow may extend several millimetres into the sediment. In consequence, the oxygen gradient does not follow the parabolic shape predicted by eq. [1.1], but has a more sigmoid shape.

With respect to the extension of the aerobic zone, lake sediments do not differ from marine ones. The sediments of eutrophic lakes generally become anaerobic within 1–2 mm (Finlay 1980; see also Fig. 4.13) unless, of course, they are covered by an anaerobic water column, while the aerobic zone is thicker in less productive lakes.

1.3.3 Microbial mats

The term 'microbial mats' describes a variety of microbial communities which grow on sediments or on solid surfaces; their matrix is formed by filamentous bacteria (Cohen and Rosenberg 1989). These may be cyanobacteria, non-oxygenic phototrophic bacteria, chemolithotrophic sulphur bacteria or heterotrophic forms. Typically many different types of bacteria show characteristic vertical zonation patterns in a mat (Section 4.1). Stromatolites belong to this category of microbial communities as do the patches of sulphur bacteria on sediment surfaces discussed above, and the biofilms in the trickling filters used for sewage treatment.

Microbial mats are typically characterized by extremely steep oxygen gradients (Kuenen *et al.* 1986; Revsbech *et al.* 1983). This is because they have very high biological activity since the dense matrix and excretion of microbial polysaccharides prevent any advective transport of solutes. Cyanobacterial mats exposed to light are oxic down to about 5 mm, and closer to the surface they are typically supersaturated with O_2; but in the dark they become anoxic and reducing up to the surface (Fig. 1.9).

1.3.4 Anaerobic microniches

Soils are basically aerobic environments since they contain an air phase and the diffusivity of gases is much greater than that of solutes. However, strict anaerobes (such as clostridia) can be isolated from soils and anaerobic

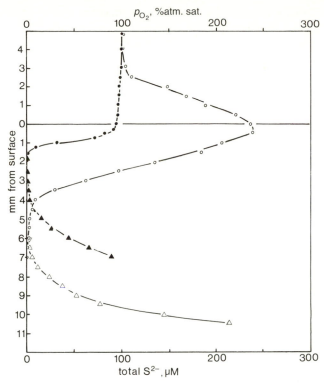

Fig. 1.9 Total S^{2-} (triangles) and p_O (circles) in the light (open symbols) and after 2 hours in darkness (filled symbols) in a sediment with a cyanobacterial/purple sulphur bacterial mat from Nivå Bay (Denmark) measured with microelectrodes. Oxygenic photosynthesis creates supersaturation in the upper 2–3 mm and makes the oxic and sulphidic zones migrate almost 4 mm downwards.

processes such as methanogenesis can be measured in normal soils. The reason for this is that water-saturated soil particles and aggregates can develop anaerobic interiors even when they are surrounded by air (see Section 1.1). The minimum diameter of such particles must be about 3 mm (Nedwell and Gray 1987). (When soils become water-logged, however, they behave like aquatic sediments and rapidly become anaerobic throughout.)

Detrital particles in aquatic environments may also have anaerobic interiors although they are submersed in oxic water. Jørgensen (1977*a*) found that sulphate reduction and sulphide oxidation may take place simultaneously at the same depth in the superficial layer of marine sediments. This was due to anaerobic and reducing conditions inside detrital particles (in particular faecal pellets of invertebrates). A smaller minimum diameter (< 1 mm) was found, and it is clear that if the external p_{O_2} is low, the minimum diameter is reduced accordingly.

It has recently been debated whether 'marine snow' may harbour anaerobic biota. Marine snow is the collection of suspended particles, consisting of various aggregated detrital material (diatom frustules, copepod faecal pellets, and the remains of gelatinous plankton organisms, all held together with microbially excreted mucus, and colonized by living microorganisms) which form in plankton. Although these particles may reach sizes ranging from 1 mm to 1 cm in diameter, they are porous and it is not evident that anaerobic conditions can develop (Alldredge and Cohen 1987). However, Bianchi *et al.* (1992) have recently isolated anaerobic bacteria and demonstrated methanogenesis in such sinking particulate material from plankton. It is conceivable that this phenomenon may be especially important in the oxygen minimum layer of oceanic water where the p_{O_2} of the ambient water may be very low (Shanks and Reeder 1993).

1.3.5 *Anaerobic habitats within animals*

The hindgut of many animals contains undigested organic material as well as a high density of bacteria, so it is not strange that it is usually anaerobic or nearly so in larger animals ranging in size from wood-eating termites to elephants. The activities of these microorganisms include a variety of fermentation processes as well as methanogensis and sulphate reduction (see Sections 2.1 and 2.2).

Many herbivorous vertebrates have independently evolved the ability to exploit this microbial activity for the utilization of structural carbohydrates (cellulose, xylan) which they are otherwise incapable of digesting. These animals have a caecum which contains anaerobic bacteria and protists. In the caecum the undigested plant debris is fermented to fatty acids which are absorbed by the hindgut. The adaptive advantage of using fermentative anaerobes is that mineralization is incomplete and that the metabolic end-products can be metabolized by the aerobic host animal. Such 'postgastric' fermentation is known from a wide variety of mammals and from some other vertebrates including perissodactyls, lagomorphs, rodents, primates, manatees and the green turtle although most have not been studied in much detail (e.g. Fenchel *et al.* 1979; Moir 1965; Murray *et al.* 1977).

Pregastric fermentation is found in ruminants and in some marsupials. The ruminants (especially cows and sheep) have been studied in great detail and the rumen probably represents one of the best understood microbial communities, especially with respect to fermenting bacteria and their mutual interactions, and the role of methanogenic bacteria (Hungate 1966, 1975; see also Sections 2.1, 2.2 and 4.1). The rumen represents about 15% of the volume of a ruminant and it can be considered as a continuous fermenter. All food (including proteins) is hydrolyzed and fermented by anaerobic microorganisms and the principle end-products are acetate, butyrate and propionate (which are absorbed and utilized by the animal) + microbial biomass (which is later digested in the abomasum or true stomach and which constitutes the protein part of the diet) + CO_2 and CH_4. Adult ruminants are thus entirely dependent on the microbial fermentation products and on microbial cells for protein.

Similar systems are also known from some invertebrate animals; most cases have been poorly studied and it is likely that there are more awaiting discovery. Wood-eating cockroaches and termites are known to harbour anaerobic microbial communities (consisting of bacteria and, in most cases, also of various anaerobic flagellates) in their hindguts and it is established that these insects are totally dependent on the microbes for the utilization of their food (Hungate 1955). Regular sea-urchins (many of which feed on macroalgae) harbour dense populations of bacteria and protozoa in parts of their gut. It is likely that this is an anaerobic microbial community which plays a role in the degradation of the food, but this has not been demonstrated experimentally. Ridder *et al.* (1985) have demonstrated that a kind of caecum in a spatangoid echinoid contains nodules covered by white sulphur bacteria. The nodules consist of parts of the detrital diet of these animals. These nodules are unquestionably anaerobic microhabitats, but the adaptive significance remains to be discovered.

2

Anaerobic prokaryotes: competition and syntrophy

2.1 Anaerobic metabolism: fermentation

The most obvious impact of fermentation in the contemporary world is on human and animal nutrition. The community of fermenting microbes living in the forestomach of herbivorous mammals is completely responsible for decomposing grass and other plant material to the fatty acids on which the host depends for growth; and wood-eating termites would find life impossible without fermenting microbes providing the same service in their hindguts. The direct exploitation by man of fermentation processes has also had a long history in most human cultures (e.g. the production of yoghurt and cheese, and the production of wine and other alcoholic beverages). Free-living fermenting microbes are also important on a global scale; especially in the anoxic strata of marine and freshwater sediments, where fermenters provide the substrates (e.g. H_2 and volatile fatty acids) used by methanogens and sulphate-reducing bacteria (producing methane and hydrogen sulphide respectively).

The organisms which live solely by fermentation are primarily microbes: most are bacteria although some are fungi or protozoa. Most spend their entire lives isolated from free oxygen: indeed, oxygen is usually toxic, and the metabolic processes they perform, especially those involved in energy generation, are relatively simple. The principal characteristic of fermenters is their uptake of preformed organic molecules, and the splitting of these into simpler organic molecules, with the conservation of some of the energy released in the reaction. The seeming simplicity of these processes, and the recognition that many are retained, with some minor modifications in all 'higher' organisms including man, has promoted the idea that fermenters are ancient, and perhaps the oldest of organisms — the original living ingredients of the 'primordial soup' (assuming the 'soup' actually contained enough organic substrates to support fermenters; see Section 1.2).

No individual fermenting microbe has the ability to completely degrade complex organic compounds (e.g. cellulose) to CO_2 and H_2, and this highlights a dramatic difference between microbial decomposition in the presence, and in the absence, of oxygen. In the aerobic environment, plant polymers are completely degraded to CO_2 and H_2O, a process that can be achieved by individual aerobic microbes. But in the absence of oxygen, complete degradation to CO_2 requires the concerted activity of several different types of anaerobes, and even then, some carbon may remain locked up in CH_4. A fundamental characteristic of the community of microbial fermenters is the interdependence of different metabolic types — the products or activities of one organism provide substrate, or an equable chemical environment, or both, for neighbouring microbes of different types. These patterns of interdependence are complex and overlapping in natural communities, where many different types of fermenters coexist. The niche to be occupied is one where biodegradable organic polymers enter an anoxic world (e.g. the rumen or lake sediment), where the absence of suitable inorganic electron acceptors (e.g. oxygen, sulphate or nitrate) makes respiration with any of these impossible; where the fermentation pathways of individual organisms are limited in the scope of their degradative ability, and where community structure has evolved to accommodate the complementary capabilities of constituent types (e.g. cellulose degraders, mixed acid fermenters, homoacetogens).

However, within each of these types, many species may compete for the same resources, and two discernible strategies operate to enhance their relative selective advantage. The first is to maximize the efficiency and energy yield of catabolic pathways and thus increase the potential growth rate. All fermenting organisms are forced to use pathways which yield only relatively small amounts of energy, but specific metabolic innovations (e.g. releasing H_2-gas) maximize the yield. The major drawback with an innovation such as H_2-evolution is that it almost always requires the close presence of a H_2-consuming organism to maintain the ambient H_2-pressure at a thermodynamically favourable low level. Thus, maximizing energy yield may *require* coexistence with an unrelated organism having a complementary metabolism; and the selective advantage gained probably underpins the evolution of syntrophic associations (see Section 2.3). The alternative strategy practised by some fermenters is the production of substances which render their surroundings unsuitable (e.g. highly acid) for growth by potential competitors. In such cases the selective pressure to maximize growth rates is reduced and the metabolic pathway is, typically (e.g. lactic acid production) relatively inefficient (although the rate of ATP generation may be very high — see below).

In this section we portray some of the essential features of fermenting organisms, the processes they perform, the energy yield of their metabolic pathways, and selected competitive strategies. The case will be made that competition and syntrophy underpin the maintenance and evolution of anaerobic microbial community structure.

2.1.1 *Energy and end-products*

The chemical reactions which are fundamentally important to the conservation of energy in biological systems are oxidation–reduction (redox) reactions. One very important biological redox reaction involves nicotinamide adenine dinucleotide (NAD) — a carrier of reducing power which readily undergoes reversible redox reactions. The oxidized and reduced forms differ by two electrons and one proton:

$$NAD^+ + 2e^- + H^+ \rightarrow NADH$$

and we can see how this works in the terminal stage of a fermentation process -the reduction of pyruvate to lactate, as it occurs in the lactic acid bacteria:

$$Pyruvate + NADH + H^+ \rightarrow Lactate + NAD^+.$$

Pyruvate is reduced to lactate, NADH is oxidized to NAD^+, and the two reactions are coupled. Each of these 'couples' (and all other redox couples) has a certain tendency to accept or lose reducing power. The tendency can be measured as an electrode potential (a voltage) at a platinum electrode, and when a couple is introduced which freely gives up electrons to the electrode, a negative potential is recorded (see Section 1.1 and Fig. 1.2). At a pH of 7 and a hydrogen pressure of one atmosphere ('standard conditions'), the couple $H_2/2H^+ + 2e^-$ has a potential of -420 mV (a redox potential at these standard conditions is designated E_0'), so hydrogen has a marked tendency to lose electrons and be oxidized. At the other end of the scale, $H_2O/O_2 + 2H^+ + 2e^-$ has an E_0' of $+820$ mV, so O_2 has a very great tendency to accept electrons and be reduced. The complete reaction, with both couples combined, then becomes:

$$H_2 + \tfrac{1}{2}O_2 \rightarrow H_2O.$$

The rule is that the couple with the lower redox potential always donates its reducing equivalents to a couple with a higher redox potential. In the reduction of pyruvate to lactate, the $NADH/NAD^+$ couple has a potential of -320 mV, and the lactate/pyruvate couple one of -190 mV. By combining both reactions, electrons tend to flow from the more electronegative couple ($NADH/NAD^+$) to the less electronegative lactate/pyruvate couple (Fig. 2.1).

Electrons move in the direction in which the free energy of a reaction process decreases, so these coupled reactions result in the loss of some free energy from the system. In the case of the reduction of pyruvate to lactate,

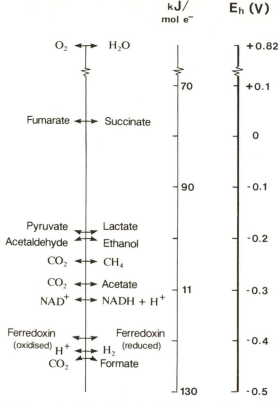

Fig. 2.1 Redox potentials (E_h) at standard conditions for some reactions in fermentation processes, expanded from the region of lowest potentials in Fig. 1.2. The standard free energy change ($\Delta G^{0\prime}$) which occurs as a pair of electrons flow from NADH to O_2 is about 220 kJ per mole (incidentally, this is close to the free energy change of 239 kJ for the non-biological oxidation of hydrogen gas — thus the biological method with NAD has evolved to optimize what is possible). Since it takes about 30.6 kJ to make a mole of ATP, it is clear that several moles can be made from the oxidation of NADH. In practice only about 3 moles are produced because of the varying efficiency of the steps in electron transfer. In the case of aerobic respiration, there are three distinct steps where the change in redox potential is very great, where much free energy is released, and where ATP can therefore be conserved. Note that the oxidation of $NADH_2$ to NAD is thermodynamically favourable when coupled with the reduction of pyruvate to lactate (which has a more positive redox potential). In contrast, coupling the oxidation of $NADH_2$ to the reduction of ferredoxin (which has a more negative potential) is thermodynamically unfavourable (unless the p_{H_2} is very low ($< 10^{-4}$ atm.); see Section 2.3).

the energy is truly lost — it is not captured and stored in an 'energy-rich' compound such as ATP because the difference between the two redox potentials is not very great. In aerobic respiration (see Section 2.2), electrons flow from NADH (-320 mV) to O_2 (+820 mV), the overall change in redox potential is great, a large amount of free energy is released, and much of this is conserved as ATP. But this option is not open to fermenting organisms: the reduction of pyruvate to lactate releases too little energy for the synthesis of ATP, which requires a differential of approximately 250 mV.

Lactate was formed because NADH had to be re-oxidized, which implies that NADH is formed earlier in the fermentation process. In catabolic reactions (those which break up large molecules such as carbohydrates, with the release of free energy) a large pool of NAD is required to pick up and carry reducing power. NAD is a large molecule and organisms cannot afford to dispose of it after a single use, so it has to be recycled. Where in the process of lactic acid production, is NADH produced and recycled?

2.1.2 *Fermentation to lactate or ethanol*

The anaerobic bacteria which produce lactic acid do so by any of three different biochemical pathways. The simplest of these, the homofermentative pathway, yields lactate alone as a final product. A six-carbon sugar (e.g. glucose) is broken down to two molecules of the three-carbon compound pyruvate. This process incorporates an oxidation reaction, where hydrogen is removed and transferred to NAD, followed by the conservation in ATP of the energy that is released (Fig. 2.2). The process following the production of pyruvate exists only for the re-oxidation of NAD: lactate acts as a sink for reducing power, and it is then excreted. So the overall process of breaking down glucose yields 2 mol lactate and 2 mol ATP and there is no net consumption of NAD. Biochemically this is one of the simplest fermentation processes.

The greater part of the lactic acid pathway — as far as the stage of pyruvate — is common to many life forms, aerobic or anaerobic. It is only after pyruvate that the pathway is substantially modified in different fermenting organisms. The smallest modification occurs in the case of alcoholic fermentation. The process has at least as much economic importance to man as lactic acid production and it is best known for being in the domain of yeasts (especially *Saccharomyces cerevisiae*). It is also produced by a variety of bacteria (e.g. *Zymomonas mobilis*), although on a global scale, alcohol production by prokaryotes is probably insignificant. Yeast has no lactate dehydrogenase, so it cannot produce lactate as a sink for the reducing power carried on NAD. What does it do? First, the carboxyl group (COO–) is removed from pyruvate as CO_2. This produces acetaldehyde, which then accepts the reducing power from NADH to form ethanol (Fig. 2.2). So, for each mol of glucose the end-products are 2 mol CO_2, and 2 mol ethanol. The ATP yield remains at 2 mol.

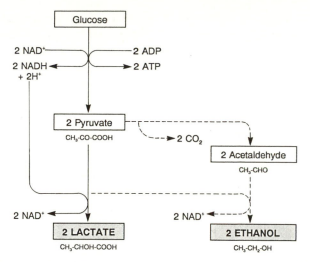

Fig. 2.2 Fermentation to lactate or ethanol. A number of intermediate reactions have been omitted.

These two simple pathways highlight the most important features of fermentations — there is an even redox balance, with no net consumption or production. If this were aerobic respiration, there would be an external acceptor (oxygen) to combine with and remove the reducing power but with fermentation, the sole vehicle for accepting and transferring reducing power is NAD, and production of NADH is balanced at some other part of the fermentation pathway by its consumption, to re-oxidize NAD.

Lactic acid bacteria are prokaryotes, and yeasts are eukaryotes, and in both cases we should not lose sight of the fact that the initial stages of energy-yielding metabolic pathways (e.g. glycolysis) in all organisms, both aerobic and anaerobic, prokaryotes and eukaryotes, are pathways which do not require the participation of oxygen. These anaerobic pathways have been retained, but they have also been adapted and extended, as organisms exploited the appearance of oxygen in the world. Even in the case of the fermenting yeast we see the retention of ancient metabolic pathways as well as the opportunistic exploitation of oxygen: metabolism is entirely fermentative in the absence of oxygen but yeasts grow aerobically when oxygen is available, producing much more cell matter for the same amount of glucose catabolized. In terms of energy yield, fermentation is inefficient, so when oxygen is available, the fermentation process is inhibited.

One of the most dramatic examples of the value of retaining simple fermentation pathways occurs in vertebrates. Hibernating turtles stop breathing and spend up to four months submerged in cold water (the record is six months; Ultsch and Jackson 1982). They continue to carry out lactic acid fermentation and they develop a severe acidosis (plasma lactate 200 mm) but they also develop an enhanced blood buffering capacity and compensatory ion exchanges which allow them to tolerate the acidosis. Goldfish on the other hand, have no means of tolerating acidosis; they produce instead ethanol which, like yeasts, they then excrete (see K. B. Storey and J. M. Storey 1990).

2.1.3 *Fermentation — definition*

At this point it is appropriate if we attempt to define what is meant by 'fermentation'. Fermentations are anaerobic energy-yielding processes in which substrates are gradually transformed by a series of balanced oxidations and reductions, in such a way that part of the substrate is oxidized at the expense of another part being reduced. No electron transport chain is involved (cf. aerobic respiration), no external electron acceptor (e.g. oxygen) is used, and ATP is conserved by substrate level phosphorlyation. With a few exceptions (e.g. Bak and Cypionka 1987) all fermentation substrates are organic molecules.

2.1.4 *Substrate level phosphorylation*

Substrate level phosphorylation (or SLP) is the principal mechanism of energy generation in fermenting organisms. In the two fermentation pathways mentioned so far, we have seen how a series of balanced redox reactions are linked to the conservation of energy in ATP. Adenosine triphosphate (ATP), the so-called 'high-energy' compound with universal distribution in the living world, has three phosphate groups, and the bonds between them have a high free energy of hydrolysis. When ATP gives up its energy, it does so by the process of hydrolysis, losing its terminal phosphate group: $ATP + H_2O \rightarrow ADP + \Pi_i$. The free energy of this hydrolysis is relatively high ($\Delta G^{0\prime} = -30.6$ kJ/mol).

In fermentation, ATP is formed by 'substrate level' phosphorylation because it is associated with redox transformations between organic compounds. We have described the oxidation of substrate coupled to the reduction of NAD: now we can see how this is linked to the conservation of energy. In the process of glycolysis, a 6-carbon sugar is first phosphorylated (this is an ATP-consuming process), then split in half to form two 3-carbon compounds (glyceraldehyde-3-phosphate), each with one phosphate group (Fig. 2.3). The first oxidation reaction involves the removal of hydrogen from glyceraldehyde-3-phosphate, and the simultaneous reduction of NAD. The free energy released from this oxidation is conserved, with the incorporation of inorganic phosphate (P_i) into a high-energy phosphate linkage, in 1,3-bisphosphoglycerate. The 3-carbon compound

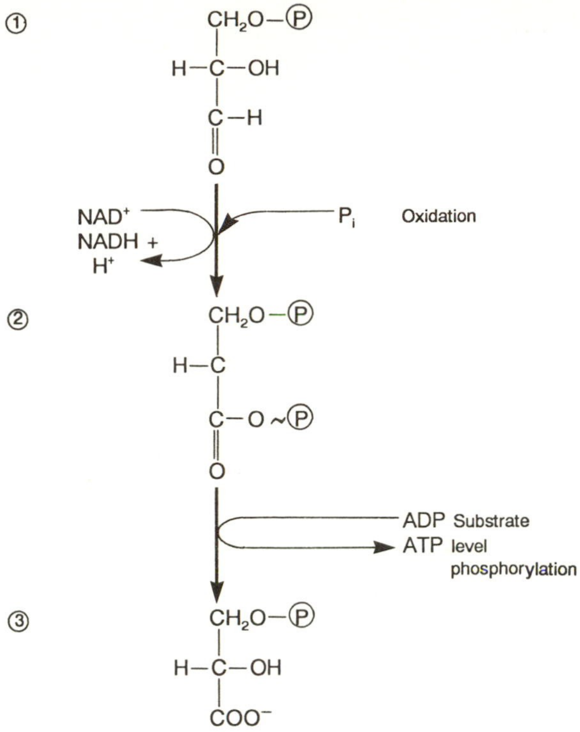

① CH$_2$O—Ⓟ

H—C—OH

C—H
‖
O

NAD$^+$ P$_i$ Oxidation
NADH +
H$^+$

② CH$_2$O—Ⓟ

H—C

C—O ~Ⓟ
‖
O

ADP Substrate
ATP level
        phosphorylation

③ CH$_2$O—Ⓟ

H—C—OH

COO$^-$

Fig. 2.3 Substrate level phosphorylation. 1, glyceraldehyde-3-phosphate; 2, 1,3-bisphosphoglycerate; 3, 3-phosphoglycerate.

now has two phosphate groups, and one of these has a high-energy linkage. In the following step, this high-energy phosphate is transferred to ADP to give ATP, gained by substrate level phosphorylation. One phosphate group remains, which is later converted to the high-energy linkage in phosphoenolpyruvate (PEP). Pyruvate kinase then cleaves this last phosphate linkage to form the second ATP by SLP. The pathways leading from *each* 3-carbon sugar yield 2 ATP. Subtracting the 2 ATP invested in the initial phosporylation of the 6-carbon sugar, the net yield of the whole process is thus 2 ATP.

2.1.5 *Acetyl CoA and mixed acid fermentations*

In the pathways examined so far, we have seen how two SLPs produce a net gain of 2 ATP/hexose. But fermentation pathways are known to yield up to 4 ATP/glucose. How is this achieved? There are two ways: one involves the incorporation of another SLP (acetyl phosphate → acetate, or butyryl-

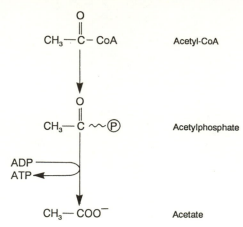

Fig. 2.4 Substrate level phosphorylation producing acetyl phosphate, with the subsequent conservation of energy in ATP and the production of acetate.

phosphate → butyrate). The other involves electron transport to an organic electron acceptor (fumarate), which, being closer to the respiratory pathways, is considered in Section 2.2.

The new SLP depends on the production of Acetyl-Coenzyme A, into which inorganic phosphate is incorporated to create the energy-rich bond in acetyl phosphate. The energy of this bond is then conserved in ATP by the agency of acetate kinase (Fig. 2.4). With the formation of acetyl-CoA, many new fermentation pathways become possible and a greater variety of end-products can be formed, even by a single organism. The mixed acid fermentation is typical, and characteristic of the enteric bacteria (e.g. *Escherichia coli*). Many of these bacteria can live in either aerobic or anaerobic environments and they have two enzyme systems for breaking down pyruvate to acetyl-CoA. Aerobically, they use the pyruvate dehydrogenase multienzyme complex which is typical of aerobic organisms generally, but anaerobically, this enzyme complex is inhibited and replaced by pyruvate-formate lyase which carries out the conversion of pyruvate (CH_3–CO–COOH) to acetyl-CoA (CH_3–CO–CoA) and formate (HCOOH). The reason is clear: the aerobic pathway through acetyl-CoA would lead to excessive production of reduced NAD (pyruvate + NAD^+ + CoA–H → acetyl-CoA + NADH + CO_2), but none is produced with the formation of formate.

Formate is then broken down (Fig. 2.5), with the help of formate-hydrogen lyase (a complex of formate dehydrogenase, a series of electron carriers and an hydrogenase) to H_2 and CO_2, and acetyl-CoA can be oxidized, eventually to acetate, with the conservation of one additional ATP (Fig. 2.6). In some respects this pathway resembles the nitrate reductase

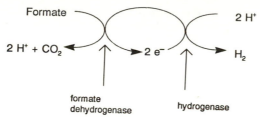

Fig. 2.5 The cleavage of formate.

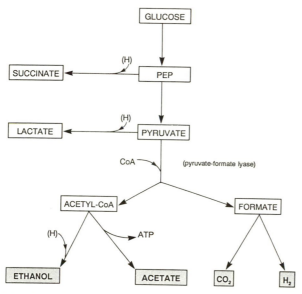

Fig. 2.6 Mixed acid fermentation, typical of the Enterobacteria, producing variable quantities of acids (succinate, lactate, acetate), ethanol, CO_2 and H_2. (H) denotes reducing power, usually in the form of NADH; PEP is phosphoenolpyruvate.

respiratory pathway. The latter also has a formate dehydrogenase and in both cases the elements selenium and molybdenum are incorporated into the enzyme. The pathways differ significantly however, in so far as nitrate respiration proceeds with the help of cytochromes as electron carriers.

Bacteria which carry out mixed acid fermentations are remarkably flexible in their response to environmental conditions (e.g. significant changes in external pH). They can modify their fermentation pathways and thus the quantities of the various end-products they produce (Table 2.1). Acetate is rarely the only end-product, reducing power is dumped at various places, and other products, especially succinate, lactate and ethanol, are formed in variable amounts. Some bacteria, e.g. *Shigella*, do not have formate-hydro-

Table 2.1. Typical products of the mixed acid fermentation in *Escherichia coli*, and the H_2-evolving fermentation in *Clostridium butyricum*. Quantities are mol/100 mol glucose.

Products	E. coli	C. butyricum
Succinate COOH–CH$_2$–CH$_2$–COOH	20	0
Lactate CH$_3$–CHOH–COOH	82	0
Formate HCOOH	2	0
Ethanol CH$_3$–CH$_2$OH	46	0
Acetate CH$_3$–COOH	40	42
Butyrate CH$_3$–CH$_2$–CH$_2$–COOH	0	76
CO$_2$	44	188
H$_2$	43	235

gen lyase, so formate is the final product and is excreted (and used by other anaerobes, e.g. methanogens). The ATP yield of a mixed acid fermentation lies between 2 and 2.5 ATP/glucose. It is increased because of the extra ATP from acetate formation. How can it be increased even further?

2.1.6 *Clostridial type fermentation*

Clostridium species grow only under anaerobic conditions and only at neutral or alkaline pH so their growth can be completely inhibited by the acid products of other fermenters (this is particularly relevant to the preservation of such foodstuffs as sauerkraut, as several *Clostridium* species, e.g. *C. botulinum,* are pathogenic to man). Metabolically they are very diverse: different species can use polysaccharides or proteins, and some (e.g. *C. pasteurianum*) can fix molecular nitrogen. They produce a wide range of fermentation products (butyrate, acetate, CO_2, ethanol, H_2, formate, succinate, and lactate).

Anaerobic bacteria have two ways to oxidize pyruvate to acetyl-CoA. We have seen one of these — the pyruvate–formate lyase system of the enteric bacteria. The other mechanism involves the iron-sulphur protein ferredoxin as an electron carrier: it is characteristic of anaerobic bacteria in the genus *Clostridium* (hence 'clostridial type' fermentation), and it functions in consort with the enzyme pyruvate-ferredoxin oxidoreductase. Similar iron-

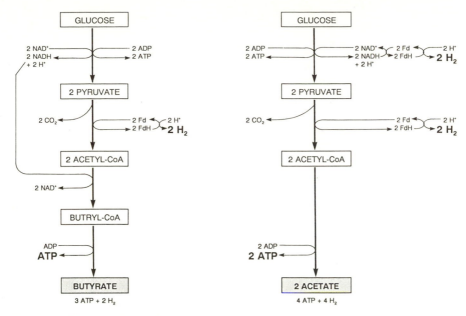

Fig. 2.7 Fermentation to butyrate or acetate.

sulphur (FeS) proteins are involved in transferring reducing power in the respiratory electron transport chains of anaerobes and aerobes, where we shall meet them again.

The ferredoxin-coupled oxidation of pyruvate proceeds as follows:

$$\text{Pyruvate} + \text{Fd}_{ox} \rightarrow \text{Fd}_{red} + \text{Acetyl-CoA} + \text{CO}_2.$$

This is an exergonic reaction ($\Delta G^{0'} = -19.2 \text{ kJ/mol}$). Moreover, ferredoxin has a very low redox potential ($E_0' = -0.41 \text{ mV}$) so it is easily oxidized, with the help of hydrogenase, to release hydrogen as gas ($2\text{H}^+ + 2\text{Fd}_{red} \rightarrow 2\text{Fd}_{ox} + \text{H}_2$). The overall process of the oxidation of pyruvate through the excretion of H_2 is not strongly dependent on $p\text{H}_2$ and it can therefore proceed in virtually any anoxic environment. The greater the capacity of an anaerobic organism to release molecular hydrogen, the less will be the need for organic hydrogen acceptors, and the energy-yielding efficiency of the whole process will be increased. Thus, in butyrate fermentation (e.g. by *Clostridium butyricum*) — some reducing power is dumped on butyrate, but some molecular hydrogen is also produced:

$$\text{Glucose} \rightarrow \text{Butyrate} + 2\text{CO}_2 + 2\text{H}_2$$

and the process yields 3 ATP/glucose (Fig. 2.7). The energy yield is limited however because it is still necessary to re-oxidize some NADH by dumping

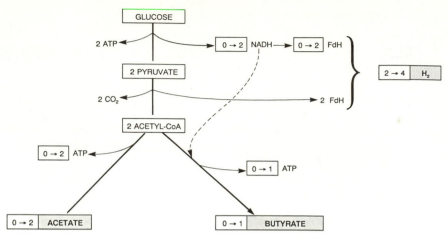

Fig. 2.8 An outline of the branched fermentation pathway in *Clostridium pasteurianum*, which produces variable quantities of acetate, butyrate, H_2 and a net yield of 3 to 4 ATP/hexose (based on information in Thauer *et al.* 1977).

on organic compounds. If this reducing power too could be released as molecular H_2, then a more oxidized end product (acetate) could be formed and more ATP could be produced. The enzyme which does this job is NADH-ferredoxin oxidoreductase. Reducing power is transferred from NADH to ferredoxin, which in turn is oxidized with the liberation of H_2. The only problem with the reaction is that it is endergonic at standard conditions:

$$NADH + Fd_{ox} \rightarrow Fd_{red} + NAD^+ + H^+ \qquad (\Delta G^{0\prime} = + 18.8 \text{ kJ/mol})$$

so the process (and the subsequent re-oxidation of reduced ferredoxin through H_2-production) cannot proceed unless the H_2 pressure is very low. But if it is kept below about 10^{-4} atm., all NADH can be oxidized, acetate can be produced as the sole end-product, and a total of 4 ATP/hexose is produced.

In practice, a more common and flexible solution is a branched fermentation, producing 3 to 4 ATP, with both butyrate and acetate as end-products. If acetate alone is produced, the net yield will be 4 ATP/hexose with the production of 4 H_2; and if the fermentation is diverted exclusively to butyrate production, the yield will be 3 ATP. A balance between these fermentation routes is maintained mainly through stimulation and inhibition of the NADH:ferredoxin oxidoreductase. Thauer *et al.* (1977) provide the example of 3.3 ATP/glucose with end-products 0.6 acetate, 0.7 butyrate, 2 CO_2 and 2.6 H_2 (Fig. 2.8).

2.1.7 *Propionate fermentation*

While there are obvious energetic advantages in channelling as much as possible of the fermentation pathway to acetate, it is rarely possible that this can be the sole end product and various other compounds must be used as sinks for reducing power. Another important branch pathway is to propionate — characteristic of the genus *Propionibacterium* — bacteria whose activity has only rarely been studied in natural anoxic habitats (e.g. Sørensen *et al.* 1981). One species (*P. acnes*) is probably responsible for the inflammation of hair follicles which takes the same name. Others give the characteristic flavour to Swiss cheeses. But perhaps the most important economically are those which live in the rumen of cattle and sheep: the propionibacteria take lactate, a major fermentation product, and ferment it further to propionate and acetate. They get one ATP from this, which probably accounts for the exploitation by these bacteria of the lactate-rich niche in the rumen (Fig. 2.9). These bacteria may just as well produce succinate (as this is the stage at which 2H is accepted) and indeed some bacteria do this. In the rumen, succinate is the major fermentation product of various *Bacteroides* species (Wolin and Miller 1989).

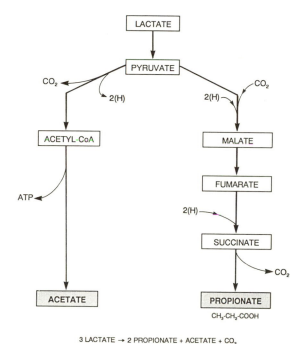

$$3 \text{ LACTATE} \rightarrow 2 \text{ PROPIONATE} + \text{ACETATE} + CO_2$$

Fig. 2.9 Fermentation of lactate to propionate and acetate.

2.1.8 *Acetogenic bacteria*

We have already seen some of the ways in which fermentation processes lead towards the production of acetate as an end-product, and several other pathways exist. Some, e.g. the Stickland reaction (the fermentation of pairs of amino acids, one of which is oxidized while the other is reduced) are probably not quantitatively important, except where much protein is being degraded:

$$\text{Alanine} \quad + \quad \text{2 Glycine} \rightarrow \text{3 Acetate} + 3NH_3 + CO_2.$$
$$\text{(CH}_3\text{CHNH}_2\text{COOH)} \quad \text{(CH}_2\text{NH}_2\text{COOH)}$$

There are however two other groups of acetate-producing fermenters. They each produce acetate by quite different methods, but they are both brought together under the umbrella of the term acetogens. The H_2-producing acetogens ferment alcohols and acids to acetate and H_2. But they can do this only if they overcome a significant thermodynamic barrier: the H_2-evolving reactions involved have a negative free energy only if the pH_2 is suitably low (i.e. $< 10^{-4}$ atm.). Thus this metabolism is possible only in natural communities, or cultures where a hydrogenotroph (e.g. a methanogen or sulphate reducer) is present. Many of these bacteria have been discovered only fairly recently and their names often reflect the obligatory nature of their syntrophic associations, e.g. the butyrate-fermenting *Syntrophomonas wolfei* (McInerney *et al.* 1981) and the propionate-fermenting *Syntrophobacter wolinii* (Boone and Bryant 1980), both of which produce acetate and H_2 as end-products.

The other group of acetogens are the homoacetogens, e.g. *Clostridium thermoaceticum* and *Acetobacterium woodii* which grow heterotrophically (when supplied with organic compounds) or autotrophically (when supplied with CO_2 or CO). Heterotrophically, organic substrates are fermented and reducing power is dumped on two molecules of CO_2, which are combined to form acetate:

$$2CO_2 + 8H^+ + 8e^- \rightarrow CH_3COOH + 2H_2O$$

whereas autotrophic cells use molecular hydrogen (H_2) as reductant for the fixation of CO_2 into acetate:

$$2CO_2 + 4H_2 \rightarrow CH_3COOH + 2H_2O.$$

The combination of two CO_2, or CO_2 and CO to form acetate is one of the simplest ways to make precursors for biosynthesis. Because of its simplicity it has been suggested that the pathway may be ancient, and perhaps one of the earliest autotrophic mechanisms of carbon fixation (Wood and Ljungdahl 1991). In this respect, it is interesting that the use of CO_2 as an electron sink is very similar to that used by the methanogens, where the electron sink is the reduction of CO_2 to methane:

$$CO_2 + 8H^+ + 8e^- \rightarrow CH_4 + 2H_2O.$$

When they are fermenting sugars to acetate, homoacetogens can produce up to 4 ATP/hexose by substrate level phosphorylation. But when acetogens grow autotrophically, the acetyl-CoA which is formed is channelled towards biosynthesis, rather than ATP-yielding SLPs. This has prompted some thorough speculation about alternative methods of ATP generation in acetogens (Dolfing 1988; Wood and Ljungdahl 1991). Homoacetogens may be able to establish a proton gradient across the cytoplasmic membrane and this could be linked to the synthesis of ATP; implying that these organisms may, in fact, be degenerate respirers. This would be consistent with the existence of cytochrome *b*, hydrogenase and NADH-ferredoxin oxidoreductase in the cell membranes of acetogens. It would also explain their unusually high growth rates (Andreesen *et al.* 1973).

The process of H_2/CO_2 acetogenesis has some ecological and global significance. It is responsible for an estimated daily production of 2.3 million tons of acetate in the hindgut of termites (Breznak and Kane 1990).

2.1.9 *What competes with fermenters?*

This question is most easily tackled by looking for the characteristics which unify fermenters and distinguish them from other groups of microorganisms. Fermenters all live in environments which are usually dark, where dissolved oxygen is undetectable, and where the redox potential is negative. They are heterotrophs (i.e. they obtain the energy they need from pre-formed organic compounds), and they acquire these compounds by diffusive transport and active uptake from the surrounding medium. If they were autotrophs, they would take up inorganic carbon (e.g. CO_2) and assimilate it into new biomass. But the biosynthesis of organic matter from CO_2 requires an input of energy, and in the environment occupied by fermenters, no suitable alternative source of energy is available. Photoautotrophs absorb light energy and convert it into a form (ATP) that is usable in biosynthesis. Chemoautotrophs obtain energy from the oxidation of inorganic compounds (e.g. Fe^{2+}, NO_3^-), usually with the help of oxygen. But, living in a dark anoxic world, and denied the ability to oxidize inorganic compounds, fermenters are limited to the anaerobic transformation of organic compounds. Thus fermenters live in a niche where they are unlikely to find themselves in serious competition with either autotrophs or aerobic heterotrophs.

Fermenters will, however, compete with each other. A diversity of species live in most anoxic habitats, and many of these have overlapping requirements. At all stages in the decomposition process there will be bacteria present which compete with each other — whether it be the pioneers in the degradation process, those that break down the cellulose and other plant structural polymers, or those that compete for volatile fatty acids or hydrogen. How do fermenters compete with each other?

It has to be said that there is very little information available from natural communities. Most of what we know comes from studies of cultures in the laboratory (especially in chemostats), and from the more accessible habitats in which fermentation dominates (e.g. the rumen, anaerobic digestors, or the hindgut of termites). Nevertheless, several key factors in the competitive strategies between fermenters may be gleaned, and these may be generally applicable in anaerobic communities. We will consider just three factors: acidity and other inhibitors, competition for substrates, and energetic efficiency.

2.1.10 *Acidity and other inhibitors*

The unpredictable business of manufacturing alcoholic beverages in France in the 1860s provided Pasteur with the first financed research project into the excretion products of fermenting microorganisms. He discovered that lactic acid produced by a contaminant would acidify the fermentation vessel to such an extent that the production of ethanol by yeasts was inhibited. If he added chalk, the medium was neutralized and alcoholic fermentation was resurrected. The lactic acid producer secured its niche by ensuring that the medium would not support growth of potential competitors. And when favourable conditions were returned to the yeast, it used the same device — producing ethanol in sufficient amounts to inhibit the growth of the lactate producers (high concentrations of ethanol inhibit glycolysis in most anaerobic prokaryotes — see Strohhäcker *et al.* 1993). This characteristic behaviour of acid-producing fermenters is well understood and it is widely exploited by man as a preservative in the foodstuffs industry (e.g. sauerkraut and yoghurt). A similar phenomenon occurs in some natural habitats (e.g. marine sediments, see Dando *et al.* 1993) as well as anaerobic digestors and landfill sites, where over-production of acid inhibits the methanogens which would otherwise complete the decomposition process (with the production of methane). These acidified habitats are often said to be 'soured'.

There is also much evidence that differences in acid tolerance are relevant to the outcome of competition in some man-made micro-habitats. Dental caries is directly related to the production of acids by fermenters, especially lactic acid producers living in tooth plaque. Many different anaerobes live in plaque, but some, notably the lactate producers, are particularly well-adapted (they are relatively oxygen tolerant) and they find their own spatial niches. *Lactobacillus casei* is found in the deepest fissures (Bowden and Hamilton 1987), and this may be related to its pronounced acid tolerance.

It still performs glycolysis at an ambient pH of 3.1 — unlike several *Streptococcus* species which might otherwise be successful competitors (Sturr and Marquis 1992).

It might be assumed that the typical acid production by fermenting organisms would inevitably lead to acidification of anoxic habitats, but this is not so. First, the acid products are usually consumed by acetogens, sulphate reducers and methanogens; and secondly, a more typical response of a fermenting organism is to modify the nature of the excreted end-products. During the course of its growth in culture, *Clostridium acetobutylicum* initially releases butyrate, but when the pH falls below 4.5, new enzymes are synthesized which enable the formation of neutral end-products such as acetone and butanol (see Schlegel 1986). This type of self-regulation is certainly too slow in its response for the ruminant which is completely dependent on the fermentation products in its rumen. These animals have a further insurance policy against acidification — the saliva they secrete almost continuously, contains various *p*H buffers (including 0.1 M bicarbonate; Hungate 1975) and it is sufficiently alkaline to neutralize slight excesses of fermentative acid production in the rumen.

2.1.11 *Acetogenesis versus methanogenesis*

In anoxic habitats which are low in sulphate and other terminal electron acceptors, H_2/CO_2 methanogenesis is the usual final electron sink. Lovely and Klug (1983) found that acetogenesis accounted for a maximum of 5% of the total combined H_2/CO_2 acetogenesis and methanogenesis in the anoxic sediment of a eutrophic lake. But in some anoxic habitats, e.g. the termite hindgut, H_2/CO_2 acetogenesis is the predominant final process (and, incidentally, the termite benefits by subsequently absorbing and oxidizing the acetate as a source of energy; Breznak and Kane 1990). The reasons for the outcome of the choice between acetogenesis and methanogenesis are still obscure. Methanogens have a greater affinity for H_2 but the acetogens are more versatile in the substrates they can use (various sugars and organic acids) and they have the capacity for switching between autotrophy and heterotrophy. An other important factor in natural habitats may be the competitive disadvantage suffered by H_2/CO_2 methanogens in some (but not all, see Williams and Crawford, 1984) acidic environments. In the permanently anoxic sediments of Knaack Lake (Phelps and Zeikus 1984) methanogenesis is limited by mildly acid conditions (pH 6.2) and the remaining methane production occurs mainly from the splitting of acetate (acetoclastic methanogenesis). Experimental raising of the pH led to an enhanced methane production. At the lower pH, the homoacetogens apparently had a selective advantage — competing successfully with the inhibited H_2/CO_2 methanogens and H_2-producing acetogens, for common substrates such as H_2, lactate, ethanol, formate and methanol.

Jones and Simon (1985) also found a stimulation of benthic acetogenesis. The lake sediment of Blelham Tarn is anoxic only during the period of summer stratification of the water column, when the quantity of anaerobic decomposition is gradually reduced because of the limited input of organic carbon. They found a system that favoured the (autotrophic) H_2/CO_2 acetogens in their competition with H_2/CO_2-methanogens (a progressive increase in acetogen activity accompanied a decline in methane production from CO_2) and an increase in acetoclastic methanogenesis: the latter presumably responding to the supply of acetate from homoacetogens. The acetogens then became an important source of acetate, providing up to 50% of that required for acetoclastic methanogenesis.

2.1.12 *Competition for substrate — cellulose*

On a global basis, the annual photosynthetic fixation of CO_2 is in excess of 10^{11} tons of dry plant biomass (Ljungdahl and Eriksson 1985) and the structural polysaccharide cellulose accounts for about half of this. Cellulose is a major constituent of all plant cells, from unicellular algae to beech trees, and when the plant dies, the first stage in its decomposition is the attack on cellulose. The degradation of cellulose is performed principally by microorganisms, especially bacteria and fungi. This is also the case in anoxic environments.

The first step in cellulose degradation is its breakdown by enzymes (cellulases) to soluble sugars. Cellulolytic organisms characteristically maximize the usefulness of these enzymes by first attaching to the substrate (e.g. small particles of grass) then secreting cellulases, and retrieving the soluble products as soon as they are produced. This initial attachment to substrate is the first place where cellulolytic organisms compete with each other, as the different organisms all try to find attachment sites. Bhat *et al.* (1990) studied adhesion of two cellulolytic ruminal bacteria to barley straw and concluded that they each had specific adhesion mechanisms allowing attachment to specific sites. With few exceptions (e.g. the study by Gelhaye *et al.* 1993 of five cellulolytic *Clostridia* coexisting in a municipal waste digestor) there is some concensus that competition for space between cellulolytic organisms does exist, but the mechanisms (both physiological and of competitive exclusion) remain obscure.

Degradation of cellulose in the rumen is not the sole province of bacteria, although several species of cellulolytic *Ruminococcus* and *Butyrovibrio* are prominent. Two other large groups of anaerobic cellulolytic fermenting organisms are both eukaryotes: chytrid fungi and ciliated protozoa. Like the bacteria, the chytrids also secrete cellulases and then take up soluble substrates. The best evidence that they compete with bacteria is circumstantial and based on the evidence of specific chemical inhibitors being produced.

The anaerobic chytrid *Neocallimastix frontalis* lives in the rumen where it carries out an efficient hydrolysis of cellulose (Bernalier *et al.* 1992). But it also coexists with the cellulolytic bacterium *Ruminococcus flavifaciens* which produces extracellular proteins. These have not yet been identified but it is clear that they are directly responsible for inhibiting the cellulolytic activity of *N. frontalis* (Bernalier *et al.* 1993) and that the bacterium benefits as a consequence.

The cellulolytic bacteria and fungi in the rumen also compete with cellulose-degrading ciliated protozoa. The ciliates are all phagotrophic, so they can ingest the smaller particles within the finely ground mixture of plant material that enters the rumen (Williams and Coleman 1989). This gives them one clear advantage: by localizing the substrate within internal food vacuoles, and by secreting cellulases into these food vacuoles, they profit from minimal wastage of enzymes. Some of these ciliates (e.g. *Epidinium caudatum*) also secrete cellulases into the surrounding medium. This seems to function as a partial digestion of plant fragments which are then more easily ingested (Akin and Amos 1979).

It is an underlying theme of this text that competition and syntrophy are two of the principal determinants of community structure in anaerobic communities. The following example also comes within the topic of cellulose degradation and it provides a rather anecdotal illustration of both factors operating within a single microbial consortium.

Cavedon and Canale-Parola (1992) have recently clarified some of the physiological interactions between an anaerobic, cellulose-degrading, N_2-fixing bacterium (*Clostridium* strain C7) and the non-cellulolytic facultatively anaerobic bacterium *Klebsiella* (strain W1). Both bacteria live in syntrophic co-culture because: (i) the *Clostridium* provides growth substrates for the *Klebsiella* in the form of cellulose-derived soluble sugars, (ii) *Clostridium* fixes N_2 (Leschine *et al.* 1988) and provides *Klebsiella* with a source of combined nitrogen (possibly amino acids), and (iii) *Klebsiella* provides the two vitamins (biotin and *p*-aminobenzoate) for which *Clostridium* has an absolute requirement. But both bacteria depend on products of cellulolysis; they both depend on the soluble sugars which are produced from the digestion of the cellulose by *Clostridium*. Since they both benefit from the syntrophic co-culture, it is in their mutual interest to minimize competition for carbon substrate. How do they manage this?

Clostridium C7 hydrolyses cellulose to cellobiose, and produces smaller quantities of cellotriose and glucose. Both the *Clostridium* and the *Klebsiella* can use all three of these as fermentable substrates. But *Clostridium* also produces some other intermediate products of higher mass (cellotetraose, -pentaose, -hexaose; Cavedon *et al.* 1990), and only the *Clostridium* can use

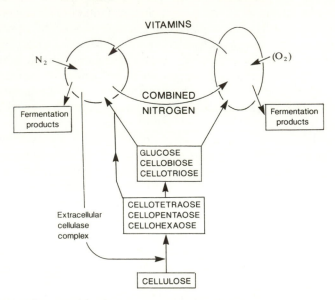

Fig. 2.10 Niche-partitioning and syntrophy by a co-culture of a cellulolytic, N_2-fixing *Clostridium* and a non-cellulolytic, facultatively anaerobic *Klebsiella* (based on information in Cavedon and Canale-Parola 1992).

these for fermentation. Thus the two partners appear to minimize competition by using different products of the cellulolysis performed by one of the partners. It is also likely that the partners differ in their affinities for the various substrates (Fig. 2.10). This phenomemon is of widespread relevance to the maintenance of a diversity of types in both aerobic and anaerobic microbial communities, for it allows bacteria to compete successfully for specific substrates within particular ranges of concentration (see below).

2.1.13 Substrate affinity

Three aspects of competitive ability, viz. 'substrate affinity', 'maintenance energy' and ATP yield, are all related to bioenergetics. For various reasons it is practical to consider them separately. Bacteria depend on active uptake mechanisms to obtain dissolved organic compounds. These mechanisms may become saturated, so that at a sufficiently high substrate concentration, the maximum uptake rate (V_m) is reached. Uptake can be adequately described by first order enzyme kinetics (Monod or Michaelis–Menten kinetics) as:

$$V = V_m[S/(K_m + S)]$$

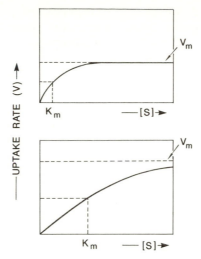

Fig. 2.11 Substrate uptake rate (V) by two hypothetical strains of bacteria as a function of substrate concentration [S]. In the upper figure, a low K_m signifies high substrate affinity. In the lower figure, maximum uptake rate (V_m) will be reached at a high substrate concentration, so the bacterium has a high substrate capacity.

where V is the uptake rate, V_m the maximum rate (i.e. uptake rate at saturation), S the substrate concentration and K_m the value for S when V is half the maximal rate. Figure 2.11 shows curves for the active uptake of a soluble organic substrate by two bacteria which differ in their uptake kinetics. In the upper panel, the bacterium becomes saturated at a relatively low concentration, the K_m is also low, and the bacterium is said to have a high substrate affinity. In the lower panel, the V_m is reached at a substrate concentration which is somewhere off scale, and the bacterium is said to have a high uptake capacity — i.e. it becomes saturated only at very high concentrations. What is the ecological significance of these two strategies? The bacterium in the lower panel will have a high uptake rate (and hence a high growth rate) at high substrate concentrations, whereas the bacterium in the upper panel will be more efficient (and thus, have a competitive advantage), at low substrate concentrations.

Within a community of fermenting microorganisms, differences in uptake kinetics probably allow several different types to coexist. In the rumen, for example, soluble substrate concentrations are often low, and sometimes variable, the rumen microbial community is species-rich, and several species are often capable of using the same substrates: thus inter-specific competition for substrates will be intense. The evolution of significant differences in substrate affinity would be one way to facilitate coexistence in this community.

Russell and Baldwin (1979*a*) looked at the growth of five types of rumen bacteria on typical rumen organic substrates (glucose, maltose, sucrose, cellobiose, xylose, lactate). They found large differences in substrate affinity within each bacterial type and between different species which used the same substrate. *Butyrivibrio fibrisolvens* had a considerably greater affinity for glucose than did *Streptococcus bovis,* whereas the reverse was true for sucrose. But uptake by *S. bovis* was still not satiated by glucose at 20 mM. Considering all coexisting species, there were sufficient differences in uptake kinetics for the variety of natural substrates to explain their coexistence.

2.1.14 *Maintenance*

In the energy budgets of microorganisms, most energy goes towards the biosynthesis of new cell material. For some non-biosynthesic processes (e.g. cell motility), energy expenditure may be trivially small (see Purcell 1977), while for others, it is at least quantifiable. The so-called 'energy of maintenance' (Pirt 1966) is the energy required for purposes such as the active uptake of substrates, ion transport, and the maintenance of ionic equilibria across membranes (see Thauer *et al.* 1977). These energy-consuming cell functions must continue even in the absence of biosynthesis, so they account for an increasing fraction of the total energy budget as the growth rate falls. This is the probable reason for the observed positive correlation between bacterial cell yield and growth rate (Pirt 1966). The maintenance energy requirement can be regarded as an integrated measure of the efficiency with which a bacterium copes with the non-biosynthetic energy demand in a particular environment. It is of interest to us here because it may have some role to play in determining the competitive fitness of different fermenting species inhabiting the same environment.

Russell and Baldwin (1979*b*) looked at maintenance energy expenditures for five rumen bacteria living in laboratory cultures. They found large differences (up to 8.5-fold), and proposed that these might influence the outcome of inter-specific competition, especially in the environment of the rumen. Bacteria with high maintenance requirements will be relatively unsuccessful (i.e. lower yield) when soluble substrates are limiting. Conversely, low maintenance requirements (and thus higher yield) will favour persistence in environments with low concentrations of soluble substrates (Fig. 2.12). Thus, at low growth rates, *Butyrivibrio fibrisolvens* has a relatively high yield because of its inherently low maintenance energy requirement, whereas *Bacteroides ruminicola* has a lower yield; the converse being the case at high growth rates. The physiological basis underlying these differences in maintenance energy requirements remains obscure, but there is some support

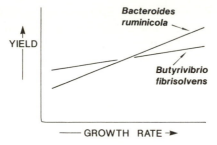

Fig. 2.12 The dependence of net cell yield on growth rate in coexisting populations of fermenting rumen bacteria may be related to maintenance energy requirements, which are lower in *Butyrivibrio fibrisolvens* (based on information in Russell and Baldwin 1979*b*).

for the idea that they do influence the outcome of competition between fermenters: *B. fibrisolvens* becomes relatively more abundant in the rumen of animals fed on a poor diet.

2.1.15 Y_{ATP}

In their studies of the growth of anaerobic bacteria, Bauchop and Elsden (1960) showed that the yield of biomass was related to the amount of ATP produced in catabolism. The term Y_{ATP} was quantified as grams (dry weight) of cells synthesized per mole of ATP produced, and the data obtained for a variety of growing anaerobic bacteria all lay close to 10 g/mol. But with the relatively greater significance of maintenance energy at low growth rates, it is clear that Y_{ATP} cannot always be constant and its value will be a function of the growth rate of the bacterium (Pirt 1966). What is the relevance of this to competition? Y_{ATP} is an index of the efficiency with which a bacterium turns simple carbon substrates into the organic macromolecules of its cell substance. This biosynthesis is the energy-consuming process, and the efficiency of this biosynthesis will of course depend on the raw materials that are available. The higher the Y_{ATP}, the higher the efficiency of conversion into new biomass and thus the higher will be the competitive fitness of the organism. Thauer *et al.* (1977) provided some figures for the maximum values of Y_{ATP} that can be expected for bacteria growing on different substrates. If glucose is the primary substrate, the energy demand of biosynthesis is relatively low ($Y_{ATP} = 27$), 3- and 2-carbon compounds are more energy-demanding ($Y_{ATP} = 13.4$ and 10 respectively), and if CO_2 is the sole carbon substrate, $Y_{ATP} = 5$. Thus it is likely that the competitive ability of bacteria in a community of fermenters will be related to their ability to acquire carbon substrates that minimize the energetic cost of synthesizing the macromolecules of new biomass. This may be particularly relevant in communities of fermenters, where the hierarchical structure may also represent a pecking order of assimilation efficiency. The cellulose degraders and primary fermenters with ready access to hexoses may then have some advantage over the autotrophic acetogens assimilating CO_2.

2.1.16 *ATP yield, substrate affinity and competitive success*

The impression is given that competitive success is dependent on maximizing the ATP yield per unit substrate and on the efficiency of biosynthesis. This is probably true in most natural communities, where substrates are usually limiting (i.e. consumed almost as fast as they are produced). But a high ATP yield/unit substrate does not automatically give a competitive advantage if the substrate is no longer limiting. This is probably a rare and transient phenomenon in the natural environment, but its occurrence in the rumen is well-documented. Fermentation pathways leading to the production of VFAs and H_2 yield more than 2 ATP/hexose whereas fermentation to lactate yields a maximum of 2 ATP. This may explain the lower abundance of lactate producers and the rapid consumption of lactate to produce volatile fatty acids (e.g. by *Clostridium propionicum*) in the rumen. But a remarkable change can occur when the substrate supply to certain lactic acid producers is substantially increased. Take the case of *Streptococcus bovis:* one important characteristic of this organism is that it has a high V_m and it is still capable of increasing uptake and metabolizing hexoses at a concentration of 20 mM. It will exercise this ability only rarely in the rumen, but it will do so when there is a sudden introduction of a large quantity of starch (e.g. in grain). Because of its high substrate capacity, the quantity of sugar *S. bovis* can ferment per unit time is much greater than that of its competitors, the quantity of ATP generated per unit time is greater and the end result is a superior growth rate (a generation time as low as 20 minutes) and massive population sizes (see Hungate 1975, 1978 and references therein). There are of course many other types of fermenters that can degrade starch in the rumen, but most have the adaptation of high substrate affinity (rather than capacity), they reach their maximum uptake rate at relatively low substrate concentrations, and the quantity of ATP they generate per unit time is significantly lower than in *S. bovis*. The final consequence is that the latter forms large quantities of lactate which cannot be consumed by other fermenters as quickly as it is produced, the rumen becomes acid and, eventually, most of the other rumen microbes are killed. But the success of *S. bovis* is short-lived: the host animal becomes seriously ill and eventually stops eating, cutting off the supply of substrate to the now acidified rumen.

2.1.17 *Structure of the whole community*

It is apparent by now that the structure of a community of fermenters depends on the co-operative action of several different functional groups of organisms. The convention is to split the community into anything from two to five functional sub-communities (e.g. Ljungdahl and Eriksson, 1985; McInerney, 1986), depending on the perspective of the writer and the type of community concerned. All the principal functional categories are il-lustrated in Fig. 2.13 which also includes the three main types of fermenting community (in freshwater and seawater, and in the digestive tracts of animals). All of these communities can use organic polymers as their starting point and in all cases the initial processes are those of hydrolyzing the polymers to soluble products, followed by fermentation to primary products (mainly H_2, CO_2 and volatile fatty acids). The two groups of acetogens then participate by increasing the proportion of acetate. The next step in the process depends on whether the community lives in the gut of an animal or is free-living.

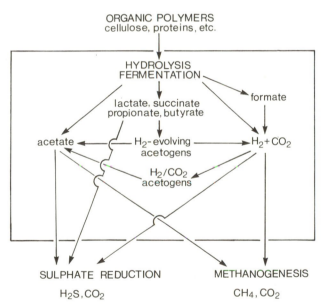

Fig. 2.13 The principal pathways for the complete anaerobic degradation of organic matter. The overall process involves the concerted action of many different types of organisms. All the fermentations lie within the box and the central roles of H_2 and acetate are clear. Sulphate reduction and methanogenesis are considered in the next section.

In the rumen or the termite hindgut, further microbially mediated decomposition is restricted to some methanogenesis, producing CH_4 which is belched or leaked to the atmosphere, but the principal end-products are acetate, propionate and butyrate which are absorbed into the bloodstream of the host animal and metabolized aerobically. In the rumen the temperature is a fairly constant 39 °C, masticated plant material is regularly injected, the fermentation products are continuously removed by the host, so the microorganisms in the community all grow relatively quickly allowing digestion and complete turnover of the rumen contents approximately once per day (Hungate 1975).

Acetate has a pivotal role to play in the process of anaerobic decomposition. It is the end-product of a number of fermentations starting from substrates with two (e.g. ethanol) or more carbon atoms (e.g. glucose), and it is produced by acetogens from 1-carbon compounds (methanol or H_2/CO_2). With so many pathways leading to the production of acetate it is obviously vital to the continued flow of carbon through the community that acetate is consumed. This is the role played by the methanogens and the sulphate reducers in, e.g. sediments, but in ruminants acetate (and other VFAs) are absorbed and utilized by the host animal. The methanogens (e.g. *Methanosarcina* spp.) consume acetate by cleaving it (in a process which is technically a fermentation) into CH_4 and CO_2 whereas the sulphate reducers (e.g. *Desulfobacter*) oxidize it to CO_2 with the reduction of sulphur compounds to H_2S. Most methanogens are also capable of coupling the oxidation of H_2 to a reduction of carbonate to produce methane. Thus in marine and freshwater environments the final products of the anaerobic decomposition of organic matter are H_2S, CH_4 and CO_2. Methanogens and sulphate reducers are considered in more detail in the next section.

2.2 Anaerobic metabolism: phototrophy, respiration and methanogenesis

We have seen how a community of different metabolic types of fermenting organisms living in an anoxic world can break down complex organic matter to produce acetate, H_2 and CO_2 as the principal end-products, and thus accomplish much of the overall catabolism. But, as we have also seen (Fig. 2.13), the complete anaerobic mineralization of organic matter yields CO_2, CH_4 and H_2S. We now turn our attention to the additional types of anaerobic metabolism which are needed to handle H_2 and acetate, and how the ultimate products of CH_4 and H_2S are produced.

The organisms concerned differ significantly from fermenters with respect to the methods used to handle reducing power. Many of these methods also provide more ATP/substrate than fermentation processes and it is possible that this improved energy efficiency has been a driving force behind evolution and diversification of these forms. Two things have been instrumental in improving the ATP yield — the use of external electron acceptors (organic and inorganic), and the transfer of energy-yielding pathways from the cytosol to cellular membranes. Phototrophs can use membrane-bound electron transport processes to convert light energy into ATP and to produce reducing power in a form they can use (NADPH) for fixing inorganic carbon (CO_2) into new cell material. In organisms which do not convert light energy, where the membrane-bound processes produce ATP from the energy delivered in reducing power (e.g. NADH), and where electrons are eventually dumped on an electron acceptor which typically is provided by, and taken up from the external world (e.g. oxygen or sulphate), the process is known as respiration. In both phototrophy and respiration, the membrane-bound processes consist of chains of molecules (e.g. Fe/S proteins, cytochromes) which typically alternate in carrying and transferring electrons and hydrogen atoms. Phototrophic and respiratory processes show great similarities with respect to the nature and juxtaposition of the carriers in the electron transport chains, and both processes are fundamentally dependent on the establishment of transmembrane proton gradients (considered below). In contrast, fermenting organisms obtain ATP solely from substrate level phosphorylations: they do not possess electron transport chains, nor do they dump reducing power on external electron acceptors or derive energy from transmembrane proton gradients.

It is, however, worth emphasizing that the boundaries between fermentation and phototrophy/respiration are diffuse. Most anaerobic phototrophs or respirers are also capable of fermentation, and in the anoxic world there are clear examples (e.g. the sulphate-reducing bacteria) of the functional integration of H_2-yielding fermentation and electron transport processes in the same organism.

One remarkable example of the arbitrariness of these categories is provided by an organism from the sulphate-reducing genus *Desulfovibrio*. Until quite recently, the utilization by anaerobes of oxidized sulphur compounds in energy generation was thought to be restricted to their consumption of these as electron acceptors for respiration. But *Desulfovibrio sulfodismutans* carries out what is essentially a fermentation of inorganic compounds — the disproportionation of oxidized sulphur compounds (Bak and Cypionka 1987). Under anoxic culture conditions, both sulphide and sulphate (the most reduced and oxidized forms of sulphur, respectively) are produced when the bacterium is grown with thiosulphate or sulphite. Growth yields are proportional to the amounts of thiosulphate or sulphite supplied, so the bacterium has an energy metabolism based on the disproportionation of sulphite or thiosulphate:

Sulphite $\rightarrow$ Sulphate + Sulphide

$4SO_3^{2-} + 2H^+ \rightarrow 3SO_4^{2-} + HS^- + H^+$

$(\Delta G^{0\prime} = -58.9$ kJ/mol sulphite)

Thiosulphate $\rightarrow$ Sulphate + Sulphide

$S_2O_3^{2-} + H_2O \rightarrow SO_4^{2-} + HS^- + H^+$

$(\Delta G^{0\prime} = -21.9$ kJ/mol thiosulphate).

The maximum energy yield is 1 ATP (from the sulphite disproportionation) which probably explains the relatively low growth rates and yield of this organism. The mechanism of energy conservation is still unclear.

2.2.1 *Fumarate respiration*

Fumarate reduction is one of the simplest respiratory processes, which is why it is considered first: the other reasons are (a) that the process has assumed a pivotal position in discussions of the evolution of fermentation into respiration, (b) because of the possibility that the biochemical pathway to fumarate has been adopted (albeit in reverse) in the citric acid cycle (which has almost ubiquitous presence in the aerobic biota, as well as having a patchy distribution in anaerobes), and (c) because fumarate reduction, whether located in the cytosol or bound to internal membranes, is a process which is extremely widely distributed in anaerobic organisms, e.g. the bacteria *Escherichia* and *Propionibacterium*, the ciliate *Tetrahymena* or the parasitic nematode *Ascaris lumbricoides* (see Section 3.5). In most or all cases, however, this type of anaerobic respiration has evolved secondarily from oxygen respiring organisms.

As shown in the preceding outline of fermentation processes (2.1), redox balance is achieved by dumping reducing power on fermentation interme- diates, e.g. the reduction of pyruvate to lactate. The disadvantage of this is that pyruvate, a compound which otherwise could be used for biosynthesis or for the generation of additional energy (e.g. pyruvate $\rightarrow$ acetyl-CoA - $\rightarrow$ acetate + ATP), is wasted. One economizing solution in the evolution of anaerobes could have been the acquisition of accessory oxidation reactions to achieve redox balance; this would seem especially plausible if there was a simultaneous liberation of some of the pyruvate for biosynthesis. The crucial first step would have been the exploitation of an accessory oxidant, most probably CO_2, which may have been available in the atmosphere at a higher partial pressure than at present. Condensation of CO_2 with pyruvate (C_3) would produce the relatively oxidized oxaloacetate (C_4). This would then be reduced to malate, converted to fumarate, then reduced further to succinate. This relatively simple chain is capable of re-oxidizing all of the NADH produced in glycolysis, thus achieving redox balance while simultaneously freeing one pyruvate for more profitable use in the cell. All of this would happen in the cytosol, as it does in some microorganisms (see Gest 1980 and Fig. 2.16a).

The evolution of this accessory oxidation chain into a membrane-bound electron transport chain coupled to phosphorylation would require the incorporation of electron carriers with suitable intermediate redox potentials. The original (soluble) fumarate reductase would have had NADH-oxidizing as well as fumarate-reducing properties, and this would have evolved into a separate NADH dehydrogenase and fumarate reductase (Jones 1985). The subsequent acquisition of menaquinone and cytochrome b and the relocation of the whole sequence within the cytoplasmic membrane would provide the essential elements of a functional respiratory chain. Once the system could also catalyse outwards directed H^+-transfer across the membrane, a proton gradient would be established which, with the development of reversible H^+-translocating ATPase complexes (the ancestor living solely by fermentation would already expel accumulated intracellular protons with ATP-consuming membrane-bound H^+-ATPases) would link electron transport, proton extrusion, the generation of a transmembrane proton gradient and the conservation of the energy in the gradient with the synthesis of ATP (Fig. 2.14). The overall process:

$$\text{Fumarate}^{2-} + \text{NADH} + H^+ \rightarrow \text{Succinate}^{2-} + \text{NAD}^+$$
$$(\Delta G^{0\prime} = -67.70 \text{ kJ/mol})$$

allows for the synthesis of at least one mole of ATP.

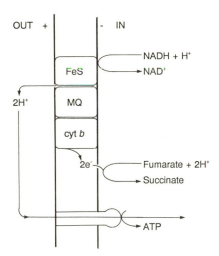

Fig. 2.14 Schematic representation of a short, membrane-bound electron transport chain with which NADH is oxidized, a transmembrane proton gradient is established (and exploited with the conservation of energy in ATP) and fumarate is the terminal electron acceptor, being reduced to succinate. FeS, iron-sulphur protein; MQ, menaquinone; cyt b, cytochrome b.

This proposed evolution of cytosolic fumarate reduction into a membrane-bound fumarate respiration has been accepted with enthusiasm (e.g. Gest 1980, 1987; Jones 1985; Behm 1991, among others). It may indeed be true that the proposed scheme adequately accounts for the evolution of this particular form of anaerobic respiration, which had its roots in a fermentation of the type described, but it should not be taken to imply that all respirations are derived from or even closely related to fumarate respiration. More importantly, the relative antiquity of fermentations to respiration processes should not be assumed. These points are dealt with in greater detail in Section 2.4.

Two groups of naturally produced methylated organic compounds can also act as anaerobic terminal electron acceptors for a variety of bacteria e.g. the non-sulphur purple genera *Rhodobacter* and *Rhodospirillum*. Trimethylamine oxide (TMAO), a nitrogen excretion product of marine fish, and dimethylamine sulphoxide (DMSO), produced especially by marine algae, are reduced to trimethylamine (TMA) and dimethyl sulphoxide (DMS) respectively; two compounds with strong and distinctive odours. The redox potential of the two couples (TMAO/TMA and DMSO/DMS) is in both cases close to 0.15 V so it is likely that they act as electron sinks for relatively short electron transport chains, which are probably curtailed at the level of cytochrome *b* (see Brock and Madigan 1991; Widdel and Hansen 1992).

2.2.2 *Chemiosmosis and electron transport*

The establishment of proton gradients across membranes is an integral part of the process of ATP synthesis in all respiring and photosynthetic bacteria, as well as in mitochondria and chloroplasts. The mechanism involved, as set out in the chemiosmotic hypothesis, was first proposed by Peter Mitchell in 1961, and there is now abundant evidence (see for example, Stryer 1988) that it is generally applicable as the mechanism which links electron transport to the synthesis of ATP.

Reducing equivalents arrive at the membrane-bound respiratory chain, usually bound to carrier molecules in the form of NADH or $FADH_2$. The purpose of the respiratory chain is to oxidize these carriers and to use the free energy that is released for the synthesis of ATP. The energy released pumps protons (H^+) across the membrane in the 'out' direction, and the matrix of the bacterium or mitochondrion becomes alkaline. An electrochemical gradient is simultaneously produced: the inside of the membrane becomes negative relative to the outside, establishing a membrane potential. The proton gradient, together with this membrane potential, constitutes the proton-motive-force (pmf). So the respiratory chain is actually a proton pump, and the proton gradient that it creates contains potential energy which can then be harnessed in ATP as protons flow back into the matrix of the cell. The route taken by protons through the inner membrane is *via* specific enzyme complexes known variously as ATPase, ATP synthase, H^+-ATPase, or F_0F_1ATPase. Within these complexes, protons flow into a zone of lower H^+-concentration, thus energy is released, and conserved as ATP.

Respiratory chains are very efficient because they provide a method of oxidizing reduced compounds such as NADH in a way which does not waste substrate such as pyruvate in soaking up reducing power. The greater the difference in redox potential between the electron donor (in this case, NADH) and the acceptor, the greater will be the free energy change and the greater the potential energy conservation. Aerobic respiration, in which O_2 is the terminal electron acceptor ($2e^- + 2H^+ + \frac{1}{2}O_2 \rightarrow H_2O$), is undoubtedly the pinnacle of achievement for a respiratory process. The redox potential difference between NADH and O_2 is very great ($E_0' = -0.32$ V and $+0.82$ V respectively) and a relatively large amount of energy is released in the oxidation ($\frac{1}{2}O_2 + NADH + H^+ \rightleftharpoons H_2O + NAD^+$; $\Delta G^{0'} = -220$ kJ/mol). The large change in redox potential over the whole respiratory chain, and the insertion during the course of evolution of suitable electron carriers, has provided three sites where electron transfer is sufficiently exergonic to drive the synthesis of ATP ($\Delta G^{0'} = -31$ kJ/mol). Re-oxidizing a single NADH with the help of such a chain yields roughly 3 ATP. Intermediary aerobic metabolism (and especially the citric acid cycle, see below) is also very efficient; it produces and channels enough NADH into respiration to give a final yield of 36 ATP/hexose. This performance cannot be attained by any form of anaerobic respiration because (a) the redox potential difference to the terminal electron accept is smaller (Figs 1.2 and 2.1), and (b) fewer ATP phosphorylation sites can be accommodated on the respiratory chain, and energy may be required (as in the case of SO_4^{2-}) simply to activate the electron acceptor.

2.2.3 *Sulphate respiration*

The sulphate-reducing bacteria are phylogenetically and metabolically diverse, but they all share two characteristics: a requirement for anaerobic conditions for growth (although some, e.g. *Desulfovibrio* are quite oxygen-tolerant); and the use of sulphate as an electron acceptor, reducing it to H_2S. When sulphate is lacking, they ferment (e.g. fumarate or malate to succinate, acetate, CO_2 and propionate). Some can also reduce sulphur, but on a global scale this is probably less important. Sulphate reducers carry out electron transport which involves one or more cytochromes (including cytochrome c_3 which is unique to sulphate reducers), ferredoxins and flavodoxins. Most sulphate reducers have been discovered relatively recently (mainly since the late 1970s) and some of the details concerning their energy-yielding metabolism remain obscure. The subject has been dealt with in several recent reviews (references in Widdel and Hansen 1992). Our purpose here is principally to highlight some relevant details which are either irrefutable or likely to be true and which serve to illustrate the functional role of sulphate reducers in the community of microorganisms living in natural anoxic habitats.

One thing that is certain is the confinement of sulphate reducers to the terminal stages of anaerobic decomposition. They cannot degrade natural polymers (polysaccharides, proteins or lipids), so they are dependent on the fermentation products of other anaerobic (mainly fermenting) bacteria. They can use H_2 and acetate as energy sources — something that is denied to fermenters which rely on the normal suite of substrate level phosphorylations; and through their exploitation of electron transport-coupled energy generation, they obtain higher ATP-yields from the oxidation of typical fermenter substrates such as lactate. Some (e.g. *Desulfococcus* spp.) can also completely oxidize fatty acids higher than acetate (e.g. butyrate) which, when coupled with an ability to oxidize H_2, gives them a competitive advantage over H_2-evolving acetogens which consume butyrate only with the help of a H_2-consuming syntrophic partner (see Section 2.3). The full range of energy substrates used by sulphate reducers is varied and includes formate, fumarate, propionate, butyrate and higher fatty acids up to C_{18}, alcohols and amino acids (see Hansen 1988). Although all sulphate reducers use sulphate as an electron acceptor, many can also use thiosulphate ($S_2O_3^{2-}$), or sulphur (S^0) (Schauder and Kröger 1993) and some (e.g. *Desulfovibrio*) can use nitrate (NO_3^-), which they reduce as far as ammonium (NH_4^+).

One method of classifying the sulphate reducers is to place them in physiological–ecological groupings (Widdel 1986), and if we do this, one of the more important dividing criteria is the ability to oxidize acetate. This produces two, clear groups. The hydrogen–lactate group includes the *Desulfovibrio* species which preferentially use H_2 or lactate as electron donor:

$$4H_2 + SO_4^{2-} + 2H^+ \rightarrow H_2S + 4H_2O$$

$$\text{Lactate} + SO_4^{2-} \rightarrow \text{Acetate} + H_2S$$

Growth with H_2 is accompanied by CO_2 or acetate as carbon source.

The second major group contains those sulphate reducers which preferentially carry out the complete oxidation of acetate:

$$2CH_3COO^- + 2SO_4^{2-} + 2H^+ \rightarrow 4HCO_3^- + 2H_2S$$

In this category, members of the genus *Desulfobacter* are pre-eminent. Biochemically (and probably ecologically) these two groups appear to overlap very little: their H_2- and acetate-oxidizing capacities are apparently mutually exclusive.

There is at least one clear disadvantage associated with using sulphate as a terminal electron acceptor; there is also one obvious advantage. The problem is that sulphate is the most stable form of sulphur in the oxic world. The redox potential of the sulphate/sulphite couple is extremely low (−516 mV), so sulphate is very unwilling to accept electrons. The only biological way to 'activate' it is with the investment of ATP to produce adenosine phosphosulphate (APS), and the APS/sulphite couple has a redox potential of −60 mV. The main advantage, of course, of being able to reduce sulphate is its great natural abundance, especially in seawater where the typical concentration of about 28 mM (depending on salinity) is rarely exhausted, other than locally (see Section 4.1).

It has to be stressed that the details of energy conservation in sulphate reducers are not completely understood. However, one of the better known examples is the mechanism in *Desulfovibrio*. The scheme shown in Fig. 2.15

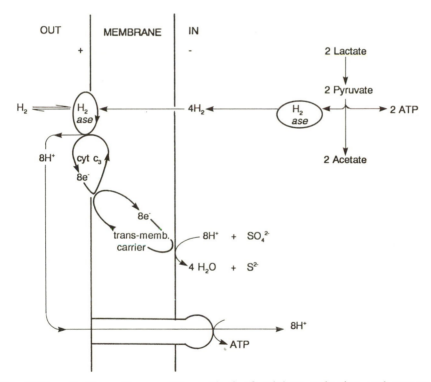

Fig. 2.15 Outline of a possible method of sulphate reduction and energy generation in *Desulfovibrio*. Molecular hydrogen is formed from the fermentation of lactate to acetate, then oxidized by a periplasmic hydrogenase which can also accept H_2 from the outside world. The electron transport chain incorporates at least one cytochrome, and a transmembrane H^+ gradient is used to generate ATP. Sulphate is used to oxidize the electrons.

is based on the mechanism proposed by Odom and Peck (1981). The organism is interesting because it performs what is really a mixture of fermentation and respiration. Within the cytoplasm, molecular H_2 is formed from the fermentation of lactate to acetate. H_2 diffuses across the membrane to be oxidized by a periplasmic hydrogenase which acts in consort with cytochrome c_3. There is a charge separation: electrons are carried to the cytoplasmic side by a transmembrane electron carrier, while the protons flow back down the proton gradient to generate ATP in the ATPase subunits. The transferred electrons are eventually disposed of in the reduction of sulphate to sulphide. In the presence of sulphate, exogenous H_2 can also be accepted by the periplasmic hydrogenase. There is some experimental evidence for such a scheme (e.g. Peck *et al.* 1987), although both it and some conceptual variants (e.g. Lupton *et al.* 1984) fail to answer such problems as failure of molecular hydrogen to inhibit growth on lactate plus sulphate (Pankhania *et al.* 1986, see also LeGall and Fauque 1988; Fauque *et al.* 1991).

The amount of ATP that can be produced by such a process is debatable (see Widdel and Hansen 1992; Nethe-Jaenchen and Thauer 1984). It must be assumed that protons are used for the import of sulphate ions and that ATP is consumed for the activation of sulphate and these theoretical considerations suggest a lower growth yield than has actually been found. So our knowledge of the mechanisms involved is obviously incomplete.

There is one further point of interest concerning this hydrogen-cycling activity within *Desulfovibrio*, because it provides an analogy, within a single organism, for inter-species hydrogen transfer in microbial consortia as outlined by Bryant *et al.* (1977), and dealt with in more detail in Section 2.3. Recall that in the case of fermentation where H_2 and acetate are the sole final products, continued hydrogen evolution is inhibited by the rising H_2-pressure. Within *Desulfovibrio*, we see a process operating which normally occurs within a consortium of two or more dissimilar microbial partners. The oxidation of lactate to acetate is barely thermodynamically favourable ($\Delta G^{0\prime} = -8$ kJ/reaction) unless it is coupled with H_2-consumption (e.g. by a methanogen in the natural environment), when it can become -144 kJ/reaction. In the case of *Desulfovibrio*, the H_2-sink is in the membrane of the same organism which produces the H_2.

2.2.4 *Acetate oxidation by sulphate reducers*

There are reasons why not all sulphate reducers oxidize acetate to CO_2. Either the initial step, the oxidation of acetate to glycolate, would require molecular oxygen or a cyclic method of oxidative decarboxylation is necessary (Baldwin and Krebs 1981; Stryer 1988). The cyclic method, which has virtually ubiquitous distribution in aerobes and which is also found in a variety of anaerobes, is the citric acid cycle, but the lactate/H_2-group of sulphate reducers do not have a citric acid cycle.

In the typical form, as found in aerobic organisms (Fig. 2.16), the acetyl unit of acetyl-CoA is combined with the 4-carbon carrier oxaloacetate to form the 6-carbon citrate. There then follows a series of oxidative decarboxylations in which the two carbon atoms introduced as acetyl-CoA are removed as CO_2, reducing power is removed as $FADH_2$ or NADH, and, in the case of the aerobic cycle, some energy (as GTP — equivalent to ATP) is conserved. The reason for a cyclic mechanism involving regeneration of a carrier molecule is clear; if oxaloacetate were not regenerated, an adult human would consume about 1 kg of it per day.

Sulphate reducers such as *Desulfobacter* use acetate as their sole energy source; they too oxidize it completely to CO_2, and they simultaneously reduce sulphate to sulphide:

$$\text{Acetate}^- + SO_4^{2-} + 3H^+ \rightarrow 2CO_2 + H_2S + 2H_2O$$
$$(\Delta G^{0\prime} = -63 \text{ kJ/reaction})$$

They have two principal ways to do this. The more usual method used by sulphate reducers (e.g. *Desulfonema*, *Desulfobacterium*) is the acetyl-CoA pathway (see Section 2.1, and below), but perhaps more interestingly, *Desulfobacter* uses a unique modification of the citric acid cycle (Fig. 2.17). While many sulphate reducers have an incomplete cycle and they cannot

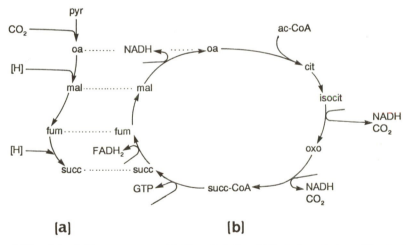

Fig. 2.16 (a) soluble accessory oxidation chain in which oxaloacetate and fumarate receive the reducing equivalents produced in glycolysis. The same sequence operates in the reverse direction in the latter half of the classical (oxidizing) citric acid cycle (b). oa, oxaloacetate; ac-CoA, acetyl-Coenzyme A; cit, citrate; isocit, isocitrate; oxo, 2-oxoglutarate; succ-CoA, succinyl-Coenzyme A; succ, succinate; fum, fumarate; mal, malate; pyr, pyruvate.

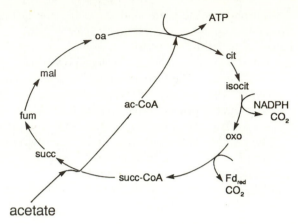

Fig. 2.17 Modification of the citric acid cycle in *Desulfobacter* showing the 'activation' of acetate and the condensation of acetyl-CoA with oxaloacetate, yielding ATP catalysed by ATP-citrate lyase.

oxidize acetate, *Desulfobacter postgatei* has all of the enzymes of the cycle. It also does two other things: first, acetate is activated to acetyl-CoA by CoA transfer from succinyl-CoA. Secondly, an enzyme, ATP-citrate lyase catalyses the condensation of acetyl-CoA and oxaloacetate to citrate. This is associated with ATP synthesis (Möller *et al.* 1987) and for each mol of acetate oxidized to $2CO_2$ by *Desulfobacter postgatei*, one mol of ATP is generated by substrate level phosphorylation. The reducing power which is released from the modified cycle (NADPH and reduced ferredoxin) is then disposed of in the reduction of sulphate to sulphide, using APS as an intermediate. Roughly 2 ATP are required to activate sulphate, so if the organism is to have any chance of growing, it really has to produce > 1 ATP by chemiosmosis. The mechanism involved and the identities of the electron carriers are poorly understood (see Möller *et al.* 1987). The reduction of sulphite (HSO_3^-/H_2S, $E_0' = -116$ mV) is feasible, with either the reduced ferredoxin ($Fd_{ox}/Fd_{red} + E_0' = -420$ mV) or with NADPH ($NADP^+/NADPH$, $E_0' = -320$ mV) and the associated free energy changes associated can probably produce about 2 ATP. The ATP yield from acetate oxidation is probably lower than from the oxidation of H_2. Where acetate is also used as carbon source, yields are about 4.8 and 7.7 g dry weight/mol acetate for *Desulfobacter postgatei* and *Desulfovibrio vulgaris*, respectively.

2.2.5 *Electron donors for sulphate reducers*

Sulphate reducers are completely dependent on the supply of fermentation products from earlier stages in decomposition, and the broad diversity of

Table 2.2. Some differences in the use of electron donors by sulphate-reducing bacteria (based on information contained in Widdel and Hansen 1992).

Genus	Electron donors				
	H_2	Acetate	Propionate	Ethanol	Lactate
Desulfovibrio	+	−	−	+	+
Desulfobacter	±	+	−	±	−
Desulfococcus	−	(+)	+	+	+
Desulfonema	±	(+)	+	−	±

+, utilized; (+), poorly utilized; ±, utilized or not utilized; −, not utilized.

forms is to some extent correlated with their differential usage of the various fermentation products which have possible use as electron donors. Table 2.2 shows the usage of electron donors by four genera (at least 13 sulphate-reducing genera are known to exist). All five electron donors are major products of fermentations, but the capacity to use each of them varies significantly between the different sulphate reducers. These differences probably have some value in facilitating coexistence of different sulphate reducers in microbial communities.

2.2.6 *Evolution of the citric acid cycle*

The citric acid cycle probably reaches its peak of functional efficiency in aerobic organisms, where all the reducing power (NADH) it produces is disposed of in the aerobic electron transport chain, which has the highest ATP yield of any known respiratory pathway. But it is also apparent that parts of the cycle (most obviously, the reductive sequence to fumarate reduction — see also Fig. 2.16) can be found in a variety of anaerobic organisms. These observations have prompted much speculation as to the origin and evolutionary development of the cycle (e.g. Gest 1987), roughly running along the lines of: (a) insertion of acetyl-CoA as the first stage of an oxidative branch, (b) reversal of the reductive sequence to succinate, and (c) joining the two arms with succinyl-CoA to form a complete, oxidative cycle. This, of course, might be challenged by other theories: we are, for example, reminded that the complete reductive cycle probably existed long before a requirement arose for feeding NADH into respiratory pathways. The complete reductive cycle is found in the autotrophic thermophilic archaebacterium *Thermoproteus neutrophilus* and the anaerobic phototroph *Chlorobium limicola* (Gest 1987; Danson and Hough 1992; Fig. 2.18), and an oxidative cycle might have arisen simply by reversal of the entire reductive cycle (see also 2.4). In addition, various methanogens have incomplete citric

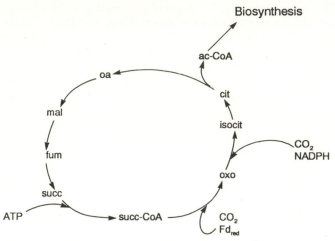

Fig. 2.18 Complete reductive cycle used for carbon fixation by the autotrophic thermophilic archaebacterium *Thermoproteus neutrophilus* and the anaerobic photo-troph *Chlorobium limicola* (based on information in Gest 1987 and Danson and Hough 1992).

acid cycles, either reductive or oxidative as far as oxoglutarate, and serving roles in biosynthesis in both cases (see Danson and Hough 1992).

2.2.7 *Other inorganic electron acceptors*

Although the use of sulphate as a terminal electron acceptor is obviously quantitatively important in the anoxic biosphere, especially in marine sediments, it is, as outlined in Sections 1.1 and 4.1, only one of several potential inorganic electron acceptors. Confining our considerations to what is thermodynamically possible in the natural world, we might expect organisms to make sequential use of electron acceptors in the order of progressively decreasing redox potentials (see Fig. 1.2). Thus, after the exhaustion of O_2, nitrate (NO_3^-) will be preferred as the next electron acceptor, followed by manganese then ferric iron (as soluble MnO_2 and $Fe(OH)_3$ respectively) and finally sulphate (SO_4^{2-}). In so far as some experimental evidence from closed anoxic systems does show that the growth and subsequent decline of nitrate-reducing bacteria precedes the appearance of sulphate reducers, the biological sequence does bear out the prediction from thermodynamics. But some other conflicting evidence does throw doubt on whether or not respiratory pathways to Mn or Fe actually exist. There is no evidence for the sequential appearance of specialist organisms which couple energy conservation with the reduction of iron or manganese, but there is evidence from studies of natural anoxic systems (e.g. Sørensen

1982; Jones *et al.* 1983; Davison and Finlay 1986) that reduced forms of manganese and iron (Mn^{2+}, Fe^{2+}) appear after the exhaustion of nitrate. Moreover, a mystery surrounds the mechanism by which Mn and Fe would become available as electron acceptors: Mn- and Fe-oxides are practically insoluble ($< 10^{-15}$ M) at neutral pH .

Many organisms are known to be capable of reducing oxides of Mn and Fe although the exact biological and chemical mechanisms involved are unknown. They could serve as a sink for excess reductant, or they could simply be reduced abiotically by the excreted products of bacterial metabolism. It is possible that they could be reduced enzymatically by respiratory enzymes in bacteria. If they act as terminal electron acceptors, the bacteria most probably involved are nitrate reducers, using their nitrate reductase system to reduce chelated Fe^{3+} or Fe oxide (see also Section 4.1). But, in the absence of any evidence that such a process is linked to energy conservation, the possibility of ferric or manganese respiration will not be considered further. The topic was reviewed by Ghiorse (1988, 1989).

2.2.8 *Nitrate respiration*

The ability to reduce nitrate as part of the process of assimilating nitrogen into nitrogenous compounds (e.g. proteins) is widespread throughout prokaryotes and in all autotrophic organisms. The enzyme involved, nitrate reductase, is soluble and it is repressed by ammonium (NH_4^+). This assimilatory process, which is quite distinct from respiratory nitrate reduction, does not concern us further; it is also relatively unimportant in anaerobic environments, which are usually net producers of NH_4^+ (see Figs 4.3 and 4.4).

Dissimilatory nitrate reduction is an energy-generating respiratory process, and the nitrate reductase is membrane-bound. With its relatively high redox potential ($NO_3^-/NO_2^- = +0.43$ V), nitrate is the most preferred alternative electron acceptor once oxygen is depleted. The microorganisms which carry out nitrate respiration are found widely in nature, e.g. soils, freshwater and seawater, and especially in aquatic sediments. All of the organisms concerned are bacteria, apart from the single known eukaryotic exception, the karyorelictid ciliate protozoon *Loxodes* (Finlay *et al.* 1983, see also Section 3.3).

The reduction of nitrate is the first step common to two broad classes of dissimilatory NO_3^- metabolism. In denitrification, the end-products are gaseous ($NO_3^- \rightarrow NO_2^- \rightarrow N_2O_{(g)}$ and $N_{2(g)}$) and the process is carried out by bacteria which in many cases can also live aerobically; nitrate reductase being derepressed (or synthesized) when oxygen is limiting. This process occurs in a wide variety of bacteria (see Tiedje 1988; Zumft 1992) including typically aerobic heterotrophs (*Pseudomonas*, *Paracoccus*), fermenters

(*Bacillus*) and purple phototrophs (*Rhodopseudomonas*) and even some halobacteria (see Zumft 1992). Nitrate reduction to ammonium ($NO_3^- \rightarrow NO_2^- \rightarrow NH_4^+$) occurs mainly in bacteria with a predominantly fermentative metabolism (e.g. *Clostridium*) although it also occurs in *Desulfovibrio*.

There are several ways in which the reduction of nitrate to ammonium is coupled to the synthesis of ATP. In *Campylobacter sputorum* and *Desulfovibrio gigas*, the oxidation of H_2 or formate is coupled to the reduction of NO_2^- to NH_4^+. However, in some *Klebsiella* species, the NADH-dependent nitrite reductase is soluble; the final product is NH_4^+, but no energy is conserved with its production. Ammonium seems to act only as an electron sink, permitting the production of more oxidized end-products, e.g. acetate, with the consequent gain in ATP from substrate level phosphorylation (as described in Section 2.1).

The basic respiratory chain incorporating nitrate reduction is shown in Fig. 2.19. The chain is a simplified version of that found in *Paracoccus denitrificans* (see Stouthamer 1988). Respiration with oxygen is also possible, with the participation of any of several alternative cytochrome oxidases (only one is shown) and nitrate reductase couples with the respiratory chain at the level of cytochrome *b*. In both cases a transmembrane proton gradient is produced and exploited for ATP synthesis in the typical fashion. The shorter chain to NO_3^- means that less ATP can be formed; the molar growth yield after switching to nitrate is typically 50% of the aerobic figure

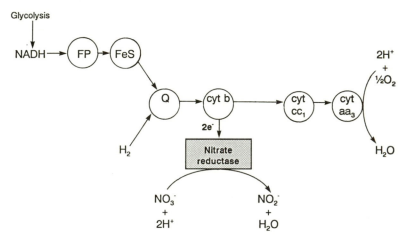

Fig. 2.19 Branched respiratory chains in a nitrate respiring bacterium (in this case, the essential features in *Paracoccus denitrificans*). Either O_2 or NO_3^- can be the terminal electron acceptor, with the branchpoint to nitrate reductase occurring at the level of cytochrome *b*.

(Stouthamer *et al.* 1980). *Paracoccus denitrificans* is one of the best studied denitrifiers — it is also rather unusual because of its additional capacity to grow with CO_2 as a carbon source and H_2 as an electron donor (i.e. it is an autotrophic nitrate respirer).

Most organisms which respire with nitrate can probably also respire with oxygen. This, together with the likelihood that nitrate was not available in sufficient quantity in the pre-oxic world suggests that anaerobic respiration to nitrate is not an ancient relic, but an example of the evolution of metabolic diversity in aerobes as part of their reversionary exploitation of anoxic habitats. We will meet the same phenomenon later in the discussion of acquired anaerobiosis in some groups of protozoa (Section 3.3), but at least in the case of denitrification, there is strong experimental evidence that nitrite reduction is a plasmid-encoded metabolic trait (Römmermann and Friedrich 1985) which can be transferred to convey the same ability to an unrelated recipient (Schneider *et al.* 1988).

There is at least one example of this acquired metabolic versatility which is interesting because it also shows that nitrate respiration does not even need to be an anaerobic process. The bacterium *Thiosphaera pantotropha* was originally isolated from a denitrifying, sulphide-oxidizing effluent treatment plant (Robertson and Kuenen 1983). It is unusually versatile, being capable of autotrophic and heterotrophic growth under both aerobic and anaerobic conditions, and able to use oxygen, nitrate, nitrite, nitrogen oxide or fumarate as terminal electron acceptors during anaerobic growth on organic carbon substrates. It can be grown autotrophically (i.e. by fixing CO_2) under aerobic and anaerobic conditions; anaerobically, using thiosulphate or sulphide as energy sources and nitrogen oxides as oxidant. One of the few things missing from its metabolic repertoire is an ability to ferment. This 'denitrifying mixotroph' has gone some way towards clarifying the phenomenon of 'aerobic denitrification' because it can simultaneously use nitrate and oxygen as terminal electron acceptors. This occurs in cultures with dissolved oxygen up to the air saturation value (Robertson and Kuenen 1984). Growth rate is higher when both nitrate and oxygen are present, and the yield with oxygen + nitrate is greater than when nitrate alone is available. The advantage to the organism (in which the nitrate reductase is constitutive and synthesized also under aerobic conditions) lies in its greater flexibility — it can rapidly switch and blend its respiratory pathways to oxygen and nitrate giving it a measure of success in a habitat such as a waste treatment plant which periodically receives a rapid injection of dissolved oxygen.

2.2.9 *Phototrophy*

Oxygen-evolving photosynthesis by green plants and algae is a process with obvious global significance: light energy is trapped by chlorophyll, transformed into chemical energy, and used to fix CO_2 into the reduced carbon of plants. Anoxygenic photosynthesis, the non-oxygen-evolving version of the process, as pursued by some anaerobic bacteria, has a less obvious impact on the natural environment although it invariably occurs wherever light penetrates to an anoxic microbial community. And where light also reaches a significant supply of reducing power (e.g. an elevated sub-surface concentration of H_2S in a stratified lake), anaerobic phototrophs are often the dominant organisms, forming dense mid-water plates where their two essential resources (light and H_2S) meet. These plates are often mono-specific, or nearly so, and when formed by purple photosynthetic bacteria (e.g. *Chromatium* spp.) they often turn the water a deep and dramatic shade of pink (e.g. Finlay *et al.* 1991). Anaerobic bacteria have an important functional role in anoxic microbial communities, because they recycle some of the products of dark anaerobic decomposition (e.g. H_2S) into new living matter. They are able to do this because they alone can harness the additional energy source in light — energy which they use to fix CO_2, with reductant (H) supplied from sources such as H_2 or H_2S, or to assimilate the acetate released by fermenters in the community (acetate is assimilated by almost all phototrophic purple bacteria; see Drews and Imhoff 1991).

2.2.10 *Anoxygenic photosynthesis — the mechanism*

The features which are common to both oxygenic and anoxygenic photo-synthesis have been known since the 1930s:

$$CO_2 + 2H_2O \rightarrow light \rightarrow (CH_2O) + O_2 + H_2O \text{ (green plants)}$$

$$CO_2 + 2H_2S \rightarrow light \rightarrow (CH_2O) + 2S + H_2O \text{ (green sulphur bacteria)}$$

In both cases, inorganic carbon (CO_2) is fixed into organic carbon (CH_2O), the source of reductant is the hydrogen in either H_2O or H_2S and the chemical energy required for this activity is derived from light energy. The sulphur that is produced anaerobically is analogous to the O_2 that is produced by the (oxygenic) photosynthesis of green plants.

Overall, these two processes are similar to each other, even if there are some notable differences in the details (e.g. differences in chlorophylls and pathways of CO_2-fixation, and the dependence of oxygenic photosynthesis on two integrated photosystems, rather than the single, cyclic photosystem of anoxygenic phototrophs). This fundamental unity is further strengthened in the similarities they jointly share with respiration — especially in their dependence on electron transport chains, and even in their use of virtually identical electron carriers (e.g. FeS proteins, ubiquinones, cytochromes). In photosynthesis, light energy is used to generate a very low potential electron donor which is then oxidized as it feeds electrons into a membrane-bound electron transport chain. This, of course, is also what happens in respiration, where a low potential electron donor (NADH) is produced, in this case by the oxidation of organic compounds. In both cases, electrons flow through the electron transport chain and a transmembrane proton gradient is generated which is used for the synthesis of ATP. In general, the structure and components of respiratory and photosynthetic electron transport chains are very similar to each other, they are each embedded in membranes, and in non-sulphur purple bacteria (some of which are capable of both anaerobic photosynthesis and aerobic respiration), the same chains are actually used for both processes.

The wavelengths of radiant energy used for photosynthesis in the biosphere all lie in the band from 200 to 1200 nm. Wavelengths greater than 1200 nm (infrared) make limited penetration into water and the energy content is too low to promote any chemical change other than the production of some heat; and wavelengths below 200 nm immediately ionize many biological molecules. Most of the light-harvesting pigments used in photosynthesis have maximum absorbance in the range 400 to 1100 nm, and in anaerobic photosynthetic bacteria, chlorophylls have evolved which absorb maximally in the band 700 to 1000 nm (see also Section 4.1).

The photochemical machinery in phototrophic bacteria has three major components: an antenna of light-harvesting pigments, a reaction centre containing at least one bacteriochlorophyll (Bchl), and an electron transport chain (Fig. 2.20). The reaction centres in photosynthetic bacteria are located within intra-cytoplasmic membranes which form tubular or vesicular structures distributed throughout most of the cytoplasm. These structures are connected to each other and (in the case of purple phototrophs) to the cytoplasmic membrane (Fig. 2.21). The reaction centres contain Bchl (of which there is a variety of types in different species) and bacteriopheophytin (Bchl minus its magnesium atom), and they are surrounded by the light-harvesting antennae molecules — mainly carotenoids which absorb at shorter wavelengths (400–550 nm), and which channel energy into Bchl. The variety of bacteriochlorophylls is striking; their diversity appears to have arisen through evolutionary tinkering with the porphyrin structure. This has produced chlorophylls with different absorption spectra which in turn allow different species to coexist as separate superimposed horizontal plates in an anoxic water body; each plate selectively filtering different wavelength bands from the underwater light spectrum (see Section 4.1).

All photosynthetic bacteria can transform light energy into a transmembrane proton gradient which is then used for the generation of ATP and for the production of NADPH. The Bchl absorbs light and becomes excited, i.e. an electron is ejected into an outer orbital. This electron is immediately accepted by another Bchl or bacteriopheophytin molecule before being transferred to a primary electron acceptor with a low redox potential —

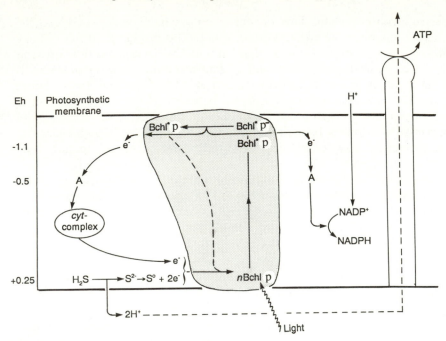

Fig. 2.20 Schematic outline of electron flow, proton translocation and ATP synthesis in the photosynthetic membrane of an anaerobic phototrophic bacterium (in this case the scheme is closest to that in a green sulphur bacterium in so far as reverse electron transport is not required to reduce the primary electron acceptor [A]). The photosynthetic reaction centre (shaded area) contains at least one bacteriochlorophyll (Bchl) which is excited (Bchl*) by light and bacteriopheophytin (P) which becomes reduced (P^-) before transferring electrons to the primary electron acceptor and thereafter into the electron transport chain (of which the cytochrome complex forms a major part) and towards the production of NADPH. The proton gradient is used for the synthesis of ATP in ATPase subunits.

usually an FeS protein; thereafter into an electron transport chain and to the synthesis of NADPH.

But this process involves a net consumption of electrons in the formation of NADPH, so an additional source of electrons is required. This is where electron donors such as H_2S and H_2 play their part, donating electrons to Bchl* and allowing it to return to its ground state. The mechanisms by which reduced sulphur compounds are employed as electron donors appear to differ between species and in most cases they are poorly or incompletely understood (see Trüper and Fischer 1982; Amesz and Knaff 1988). For example, H_2S probably transfers its electrons to oxidized cytochrome c:

$$H_2S + 2cyt^{3+} \rightarrow 2H^+ + S^0 + 2cyt^{2+}$$

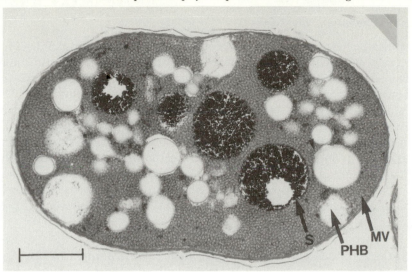

Fig. 2.21 Transmission electron micrograph of a thin section from the purple sulphur bacterium *Chromatium weissei*. The cytoplasm contains sulphur (S) granules (the product of photosynthetically mediated sulphide oxidation), and lipid-like storage inclusions consisting of poly-β-hydroxybutyrate (PHB). The remainder of the cytoplasm is almost completely filled with photosynthetic membrane vesicles (MV). The scale bar represents 1 μm (photo courtesy of K.J. Clarke, IFE).

with the involvement of sulphide:cytochrome c oxidoreductases, but the mechanism of electron transfer to Bchl is unknown. The consumption and translocation of protons in the production of NADPH, together with the splitting of H_2S, both contribute to the proton gradient (inside acid) which can then be exploited in the genetration of ATP by membrane-bound ATPase units.

Excitation by light reduces the redox potential of the reaction centre in purple bacteria from about +0.5 V to −0.7 V, whereas in green sulphur bacteria and heliobacteria it is reduced from about +0.25 V to −1.1 V. This difference between groups of phototrophs does have the consequence that the primary electron acceptor in purple bacteria is poised at a proportionately higher potential (about −0.15 V) compared to the green bacteria and heliobacteria (about −0.5 V). In the latter, this potential is sufficiently negative to reduce NADP$^+$ directly ($E_0' = -0.32$ V), whereas in purple bacteria an investment of ATP is required for the reversed electron transport needed to create a potential more negative than that of the NADPH/NADP$^+$ couple.

2.2.11 *Phenotypic diversity*

Historically, the apparent usefulness of colour as a phenotypic character has been fully exploited for the identification and classification of phototrophic bacteria. Originally there were just two groups — the purple bacteria and the green bacteria. As more characters were introduced (motility, intracellular sulphur, gas vesicles, etc.) the classification became more complex; and now, as with the classification of prokaryotes generally, it is being completely transformed by the introduction of sequence data from small subunit ribosomal RNAs (Woese 1987) and the creation of more objective evolutionary phylogenies (see Section 2.4). There is no general agreement as to the number of groups of anoxygenic phototrophic bacteria which can or should be recognized although we can tentatively list the following:

1. Purple bacteria, separated into:

    (a) purple sulphur bacteria, e.g. *Chromatium, Thiocapsa* (no gas vesicles), *Amoebobacter, Thiopedia* (with gas vesicles), and
    (b) purple non-sulphur bacteria, e.g. *Rhodospirillum, Rhodopseudomonas, Rhodobacter.*

2. Green sulphur bacteria, e.g. *Chlorobium.*

3. Green non-sulphur bacteria, e.g. *Chloroflexus* (also known as 'multicellular, filamentous, or gliding green bacteria'.

4. Heliobacteria, e.g. *Heliobacterium chlorum.*

5. Halophilic archaebacteria, e.g. *Halobacterium* spp.

Taken as a whole, the phototrophic bacteria represent many metabolic types. Green sulphur bacteria are strict anaerobes and obligate phototrophs with an absolute requirement for reduced sulphur compounds. Green non-sulphur bacteria such as *Chloroflexus* can live as anaerobic phototrophs, using H_2S and CO_2: they can also live in the dark as aerobic heterotrophs. The non-sulphur purple phototrophic bacteria are probably the most diverse group within the eubacteria (see Section 2.4). They are called 'non-sulphur' because it was thought that they could not use sulphide as an electron donor. It is now known that they can, but only at low sulphide concentrations: high concentrations are apparently toxic. Under anaerobic conditions, the non-sulphur purple bacteria are phototrophic, using H_2 or H_2S as electron donor and CO_2 as carbon source, although they can also use some organic compounds, e.g. lactate, as both electron donor and carbon source. In the dark, they can be aerobic heterotrophs, using the same electron transport chain for oxidative phosphorylation that they use for cyclic photophosphorylation in the light. *Rhodobacter capsulatus* can be photoautotrophic with H_2S or H_2 as electron donor, and with a variety of organic compounds as both electron and carbon sources, chemoautotrophic with H_2, and chemo-heterotrophic with O_2 and NO_3^-. It can also ferment organic compounds (Drews and Imhoff 1991). Purple non-sulphur bacteria have a membrane-bound linear electron transport chain and they need to synthesize only a terminal cytochrome oxidase to catalyse the reduction of O_2 to water. The chain has fewer components than the mitochondrial chain but the carriers are functionally equivalent in both. The recent discovery of such bacteria as endosymbionts of anaerobic ciliates (Fenchel and Bernard 1993*a,b*; see also Section 3.4) has stimulated renewed interest in identifying the prokaryotic ancestors of contemporary mitochondria.

Phototrophic bacteria which resemble the non-sulphur purple bacteria *Rhodomicrobium vannielii* and *Rhodopseudomonas palustris* have recently been shown to be capable of using Fe^{2+} as an electron donor for anoxygenic photosynthesis and CO_2 fixation (Widdel *et al.* 1993). At pH 7 the various Fe^{3+}/Fe^{2+} couples have redox potentials in the range -200 to $-240\,mV$, so Fe^{2+} could in principle be used as electron donor (the mid-point potential of primary electron acceptors in purple bacteria is about $-150\,mV$). If the process was quantitatively significant in the pre-oxic world, there would of course be little justification for believing that the Fe(III) in banded iron formations was exclusively the product of chemical or biological processes involving molecular O_2 as electron acceptor (see Section 1.2).

The metabolic versatility in the phototrophic non-sulphur bacteria does not seem to be shared by the purple sulphur bacteria: *Chromatium okenii*, for example, is a strictly anaerobic, obligate photoautotroph, using only H_2S or elemental sulphur as electron donor. There are also big differences between phototrophic bacteria regarding their assimilation of organic compounds. Acetate is assimilated into cell material quite easily — especially with the energy supplied through photophosphorylation, but *Chromatium* spp. appear to photoassimilate only acetate and pyruvate (Drews and Imhoff 1991). The non-sulphur purple bacteria are less discriminating. Almost all of them can assimilate acetate. *Rhodospirillum* spp. can also grow with a wide variety of organic acids (malate, fumarate, succinate) and even fatty acids and carbohydrates (Madigan 1988). Although the principle function of these organic compounds is to provide a source of carbon for biosynthesis, it is also probable that some of them, including acetate, can act as electron donors, especially if inorganic electron donors (e.g. H_2S) are unavailable.

2.2.12 *Other anoxygenic phototrophs*

Some mention should also be made of anoxygenic photosynthesis in three other groups of bacteria: the heliobacteria, the halobacteria and the cyanobacteria. Some cyanobacteria (e.g. *Oscillatoria limnetica*) are capable of anoxygenic photosynthesis with sulphide as electron donor, but as a group, the cyanobacteria are basically aerobic phototrophs with two photosystems, and chlorophyll *a* (Cohen *et al.* 1986). Relatively little is known about the recently discovered (Gest and Favinger 1983) heliobacteria (see Amesz 1991; Brock and Madigan 1991). They are strictly anaerobic phototrophs, with the unique bacteriochlorophyll *g*. They are found in anoxic soils, and they can fix N_2. There is no evidence of membrane vesicles or indeed any other cytoplasmic invaginations of a photosynthetic membrane. The primary electron acceptor has a low redox potential (-500 mV), as also found in the green sulphur bacteria. In other respects (the dependence on photoheterotrophy), they are closer to the non-sulphur purple bacteria. Sequencing of 16S ribosomal rRNA shows that they are actually more closely related to the clostridia (see Section 2.4).

The Halobacteria (e.g. *H. halobium*) are even more unusual, most obviously because they do not grow at salt concentrations lower than about 3 M (in low-sodium waters the cell wall of the organism collapses). Typically, they are heterotrophs with an aerobic electron transport chain, but they can also respire anaerobically, with nitrate or S^0. In the absence of suitable electron acceptors they can ferment carbohydrates. But, most remarkably, when exposed to light in the absence of oxygen, a purple membrane is synthesized. This contains a single photo-sensitive protein — bacteriorhodopsin — which, when illuminated, undergoes cyclic bleaching and regeneration, causing the extrusion of protons from the cell. A transmembrane proton gradient is established, and ATP is synthesized in typical ATPase units within the membrane. This light-stimulated proton pump operates without the participation of electron transport. ATP synthesized by this method is sufficient to allow slow growth in the absence of oxygen (see also Section 2.4).

2.2.13 *Methanogenesis*

The last major group of organisms involved in the anaerobic mineralization of organic matter are the methane-producing bacteria (methanogens). In common with the sulphate reducers, they are obligate anaerobes, but unlike the latter they are relatively restricted in the substrates they use. They are, for example, unable to use the volatile fatty acids butyrate and propionate which are common substrates for some sulphate reducers (e.g. *Desulfobacter* spp.). Moreover, unlike the sulphate reducers, methanogens typically lack the cytochromes and other components (e.g. quinones) of electron transport chains. At least two cytochromes are present in *Methanosarcina* spp. but their function is unclear: they may be involved as electron carriers in the disproportionation of acetate (see Whitman *et al.* 1992).

The substrates of methanogens fall into three classes:

(1) CO_2, including formate (HCOOH) and carbon monoxide (CO), with H_2 as the typical electron donor;

(2) the methyl (CH_3) groups of C-1 compounds, e.g. methanol (CH_3OH), methylamine ($CH_3NH_3^+$); and

(3) acetate (CH_3COOH).

With a few notable exceptions (see Table 2.3) most methanogens are able to use H_2 as an electron donor for the reduction of CO_2.

In the autotrophic (H_2/CO_2) methanogens, CO_2 is used as terminal oxidant in energy-yielding metabolism. This has promoted common usage of the term 'carbonate respiration' for this form of methanogenesis. It is however unwise to assume that carbonate respiration is analogous to other forms of respiration (e.g. to oxygen, sulphate or fumarate). Although it is true that less is known about the biochemical pathways of methanogenesis than of the true respiratory pathways, that which is known shows that methanogenesis is significantly different to anaerobic respiration: thus it is considered separately here. Some methanogens also carry out sulphate and sulphur reduction (see Fuchs *et al.* 1992) but the capacity to couple energy generation to methane production is exclusively a characteristic of the methanogens.

Table 2.3. Differential use of methanogenic substrates by some common genera of methanogens (adapted from Brock and Madigan 1991, with further information from Whitman *et al.* 1992). The total number of described genera of methanogens is 19.

Genus	No. of named species	Substrate		
		H_2/CO_2 or Formate	Methanol/ Methylamines	Acetate
Methanobacterium	12	+	−	−
Methanobrevibacter	3	+	−	−
Methanococcus	6	+	−	−
Methanolobus	3	−	+	−
Methanosaeta (formerly *Methanothrix*)	3–4	−	−	+
Methanocorpusculum	5	+	+	−
Methanosarcina	6	+ (Not formate)	+	+

Typical examples of energy generation with the production of methane, based on the three types of substrates mentioned above, proceed as follows.

1. The reduction of CO_2 (the autotrophic methanogens) or formate to CH_4:

    $$CO_2 + 4H_2 \rightarrow CH_4 + 2H_2O \qquad (\Delta G^{0\prime} = -136 \text{ kJ/mol } CH_4).$$
    $$4HCOOH \rightarrow CH_4 + 3CO_2 + 2H_2O \qquad (\Delta G^{0\prime} = -130 \text{ kJ/mol } CH_4).$$

2. The reduction of the methyl group (CH_3) in methanol and methylamines:

    $$4CH_3OH \rightarrow 3CH_4 + CO_2 + 2H_2O \qquad (\Delta G^{0\prime} = -105 \text{ kJ/mol } CH_4).$$

    This is equivalent to a type of fermentation, because some substrate molecules are oxidized to CO_2 while others are reduced to methane. Methanol and methylamines are derived, respectively, from pectin (Schink and Zeikus 1980) and choline (Fiebig and Gottschalk 1983).

3. The disproportionation of acetate ('acetoclastic' methanogenesis):

    $$CH_3COOH \rightarrow CH_4 + CO_2 \qquad (\Delta G^{0\prime} = -31 \text{ kJ/mol } CH_4).$$

    This requires the ATP-dependent activation of acetate to acetyl-CoA. The relatively low energy yield for this form of methanogenesis is reflected in the consistently low growth rates of methanogens which depend solely on splitting acetate: generation times under optimal conditions are typically 1–3 days (see Whitman *et al.* 1992) and as long as 9 days in *Methanosaeta* (formerly *Methanothrix*) *soehngenii* (Zehnder *et al.* 1980).

Some significant questions still remain concerning the mechanisms of energy generation in methanogens. There is no known biochemical mechanism for coupling substrate-level phosphorylation to methanogenesis, so energy generation must be by electron transport and chemiosmotic coupling, as in other forms of anaerobic respiration. But how do methanogens do this if the typical components of electron transport chains are lacking? Even if the few known cytochromes (in acetoclastic methanogens) are involved in electron transport, the nature of that involvement must be unusual: the relevant redox couples (H_2/H^+ and CO_2/CH_4) are so low (approximately −400 to −200 mV) that any electron transport chains involved would be very different to anything else known for respiratory pathways.

The discovery of a large number of unique coenzymes in methanogens and of the vital role they play in methanogenic energy-yielding metabolism, confirms the special status given to methanogens. These coenzymes include methanofuran, which participates in the initial step of H_2/CO_2 methanogenesis (the reduction of CO_2 to the formyl [O–C–H] level), and coenzyme M (CoM) which carries the methyl (CH_3) group in its final reduction to CH_4. Coenzyme F_{420} is especially interesting: it is present in large quantities in

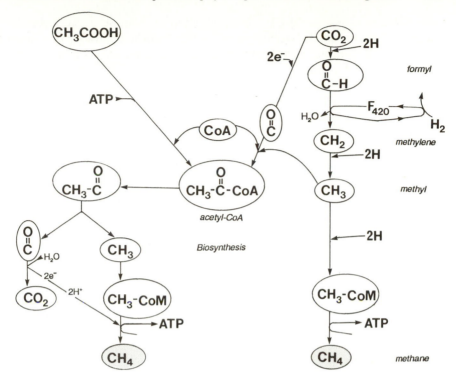

Fig. 2.22 Outline of methanogenic pathways from CO_2 and from acetate (CH_3COOH), and their integration, through the focal compound acetyl-CoA, with biosynthesis. F_{420} is the ferredoxin analogue, coenzyme F_{420}.

H_2/CO_2-methanogens and its structure is similar to the flavin coenzymes in most other (aerobic and anaerobic) organisms. When oxidized, it absorbs ultraviolet and violet light and emits fluorescence, providing a vivid signal of the presence of methanogens in mixed microbial communities (see examples in Section 3.4). The coenzyme is an electron carrier with a redox potential of -350 mV, it accepts electrons from H_2 (via hydrogenase) and its function is thus roughly the same as that of ferredoxin (see Section 2.1).

In all methanogens, the bioenergetic and biosynthetic pathways are intimately connected with each other and it is most useful to consider them together (Fig. 2.22). No methanogen has a complete citric acid cycle, neither oxidative nor reductive although partial cycles may serve biosynthetic functions (Danson and Hough 1992). Biosynthesis is typically by the acetyl-CoA pathway, which is also common to homoacetogens (see Section 2.1) and sulphate reducers. In the autotrophic (H_2/CO_2) methanogens, the methanogenic pathway leads to the production of CH_3 (methyl) groups. As these methanogens grow with CO_2 as the sole carbon source, they have no other means of acquiring CH_3-groups for biosynthesis so a proportion is diverted from the methanogenic pathway. The branched pathway producing CO from CO_2 (using carbon monoxide dehydrogenase), followed by the incorporation of coenzyme A, and CH_3 from the methanogenic pathway, satisfies the autotrophic methanogens' requirement for acetyl-CoA, the initial building block for biosynthesis.

Acetyl-CoA is also the focal point in acetoclastic methanogenesis. The methyl group is separated, and the oxidation of the residual CO to CO_2 provides electrons for the final reduction of CH_3–CoM to CH_4. Autotrophic and acetoclastic methanogens both share the same final conversion of CH_3–CoM to CH_4. This is known to be linked to the establishment of a transmembrane proton gradient, possibly involving an hydrogenase in the cell membrane of the autotrophic species (Doddema *et al.* 1979), and to the synthesis of ATP, but the details of the mechanisms involved are still unclear (see Oremland 1988; Fuchs *et al.* 1992; Wolfe 1992).

At first sight, methanogenesis from substrates other than acetate would support an energetically reasonable performance. If we assume a $\Delta G^{0\prime}$ of approximately 40 kJ/mol ATP formed from ADP, the -105 to -136kJ/mol (methanogenesis from CO_2, formate or methanol) should yield about 3 ATP (Vogels *et al.* 1988). But performance is unlikely to be so impressive in the natural environment. In habitats supporting H_2/CO_2 methanogenesis, the H_2-pressure is rarely greater than 10^{-3} atm., so $\Delta G^{0\prime}$ will be closer to -40 kJ/mol. Even if ATP synthesis worked at 100% efficiency the process is barely sufficient to produce 1 ATP (which also applies to acetoclastic methanogenesis, with a $\Delta G^{0\prime}$ of -31 kJ/mol). Considering their restricted usage of substrates, poor energetic performance and low growth rates, it is not absolutely clear how methanogens manage to make a living in the natural environment. They do however live in habitats where competition with other microorganisms is minimal (especially those in which sulphate is limiting), and their chances of success must be enhanced by their unique usage of certain energy substrates, e.g. methylamines. As we shall see (Section 3.4) they have also been uniquely successful at invading habitats such as the cytoplasm of anaerobic eukaryotes, and in retaining their hold on others (e.g. anaerobic submarine hot vents with temperatures around 100 °C; see Section 2.4).

This brief outline of non-fermentative energy metabolism has highlighted two different strategies employed by anaerobic prokaryotes. However, the boundaries separating fermentative and non-fermentative energy metabolism may be diffuse. First, the essential characteristics of respiration and phototrophy are not the electron transport chains, but the process of membrane-bound energy generation (basically, the synthesis of 1 ATP per $3H^+$ passing through an ATPase complex). In the phototroph *Halobacterium*, a light-induced transmembrane proton gradient promotes membrane-bound synthesis of ATP in ATPase complexes. This occurs without the involvement of an electron transport chain. Secondly, the heterotrophic anaerobic bacterium *Propionigenium modestum* carries out the following as its sole energy-yielding reaction:

$$succinate^{2-} + H_2O \rightarrow propionate^- + HCO_3^-$$
$$(\Delta G^{0\prime} = -20.5 \text{ kJ/reaction})$$

The bacterium is specialized for using succinate as energy substrate and is thus dependent on fumarate reduction by other anaerobes, but the energy released from the decarboxylation of succinate is insufficient to drive substrate level phosphorylation (assuming at least 31 kJ are required to synthesize 1 ATP). The bacterium solves the problem by coupling the decarboxylation of succinate to the extrusion of Na^+ ions. Thus a Na^+ gradient is established across the plasma membrane (outside higher), which is then used by a membrane bound Na^+-transporting ATPase for the synthesis of ATP. The energy gained from this process is 'modest' but it will support growth of *P. modestum* (see Schink 1992*b*).

Biochemical ingenuity in anaerobic bacteria does seem to have stretched to the limits of what is thermodynamically possible. This ingenuity has undoubtedly arisen both by evolution and by the exchange of 'useful' genes (e.g. by plasmids). The capacity of syntrophic associations of micro-organisms to break out of the thermodynamic limitations which constrain isolated organisms is discussed in the next section.

2.3 Syntrophy

One of the principal messages to emerge from our discussion of anaerobic metabolism in natural communities (Sections 2.1 and 2.2) is that the catabolism of organic matter depends on the participation of several different types of organisms carrying out different functions. In the intermediate stages we saw how more oxidized end-products (e.g. acetate) were produced; and we considered briefly how this depended on H_2-evolving reactions (e.g. the oxidation of propionate) and to consumption of H_2 by other organisms such as homoacetogens, methanogens and sulphate reducers. The process involved, so-called 'interspecies H_2-transfer' lies at the heart of the coherent function of anaerobic communities: one or more H_2-consuming species alter the metabolism of their fermenting partners towards the production of more oxidized end-products. The same process also underpins the many examples of 'syntrophy' (literally 'feeding together') in anaerobic communities, but this term is currently used in a variety of ways to refer to different types of processes. On the one hand it can be broadly defined as e.g. 'metabolites of one type of organism serving as substrates for other types of organism', but on reflection, the entire biosphere could then be said to be composed of syntrophs all playing their part in a Gaia theory of the Earth. Understanding the real importance of syntrophy for the structure and evolution of anaerobic communities requires a tighter definition of syntrophy.

Methanogenic bacteria are extremely sensitive to O_2 so they are typically found in anoxic habitats such as aquatic freshwater sediments. In addition, it is likely that they can also live in anoxic microhabitats (e.g. faecal pellets; see Alldredge and Cohen 1987) located in oxic environments, so long as they are closely associated with aerobic organisms which have a respiratory activity great enough to maintain a vanishingly low p_{O_2}. Recent experimental evidence with laboratory co-cultures of methanogens and strict aerobes (Gerritse and Gottschal 1993) shows that these two types of organisms can live together in oxic environments because the respiratory activity, and in particular the very low apparent k_m for O_2 uptake (in the nanomolar range) of some aerobes, can maintain dissolved O_2 concentrations low enough to avoid O_2-inhibition of methanogens. However, aside from the complementary O_2-dependencies of these two groups of organisms, there is no obvious way in which they interact metabolically. The interaction is limited to the presence of one partner making possible the growth of the other partner, but the association is not obligatory. The aerobes will still function if the methanogen is removed, and if the latter finds itself surrounded by anoxia, it will continue to function. In contrast, obligate syntrophy would involve some metabolic transfer or exchange between the partners, and this would be absolutely necessary for an essential metabolic reaction to proceed in at least one of the partners. The most comprehensible examples of this latter form of syntrophy involve interspecies H_2-transfer between anaerobic microorganisms. It is these associations with which we are principally concerned here.

2.3.1 *Obligate syntrophy*

The unravelling of the process of obligate syntrophy has been a protracted affair. The roots of the story probably lie with the methanogen discovered by Omelianski in 1916 (and named *Methanobacillus omelianskii*) which, it was claimed, was capable of oxidizing ethanol. But it was not until the late 1960s (Bryant *et al.* 1967) that the agent responsible was shown to be a pair of dissimilar organisms — the chemotrophic 'S-organism', and a methanogen, thereafter named *Methanobacterium bryantii* (Fig. 2.23). Neither organism on its own was capable of using ethanol as a substrate: the methanogen because it did not have the biochemical capability, and the S-organism because the process of ethanol oxidation produced H_2 which inhibited further oxidation (the S-organism cannot carry out this process at

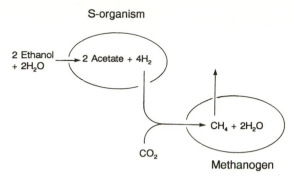

Fig. 2.23 '*Methanobacillus omelianskii*' — a syntrophic consortium of two organisms which are completely dependent on one another. $\Delta G^{0\prime}$ for the S-organism and the methanogen are, respectively, 19 kJ/2 ethanol and −131 kJ/methane, so the overall syntrophic reaction is exergonic (−112 kJ/methane).

standard conditions). The demonstration that the H_2 pressure was important, and that 'interspecies H_2-transfer' is the crucial process, was provided by Iannotti *et al.* (1973) using a co-culture of *Ruminococcus albus* and *Vibrio succinogenes*. When cultured alone, *R. albus* fermented glucose to acetate, ethanol, H_2 and CO_2, but when *V. succinogenes* was added, together with fumarate (which *R. albus* cannot reduce) the fermentation products of *R. albus* shifted to acetate and CO_2 with a simultaneous increase in cell yield. The inescapable conclusion was that the H_2 was removed by *V. succinogenes* (for fumarate reduction to succinate) and that the inefficient production of ethanol as an electron sink was circumvented. Again, both partners benefit from the increased efficiency of the overall process.

2.3.2 *Energetics*

The crucial role of the H_2-pressure becomes apparent if we take a closer look at the energetics of the oxidation of ethanol. At standard conditions, the reaction is endergonic and so unlikely to proceed:

$$\text{Ethanol} + H_2O \rightarrow \text{Acetate}^- + H^+ + 2H_2 \qquad (\Delta G^{0\prime} = 10 \text{ kJ/mol}),$$

but the free energy of the reaction is dependent on the p_{H_2} (Fig. 2.24), and at partial pressures less than about 0.15 atm., the oxidation becomes exergonic, and increasingly so as the p_{H_2} is further reduced. Butyrate and propionate oxidation are also sensitive to the H_2-pressure. At standard conditions these reactions too are endergonic:

	$\Delta G^{0\prime}$ (kJ/mol)
$\text{Butyrate}^- + 2H_2O \rightarrow 2 \text{ Acetate}^- + H^+ + 2H_2$	+48
$\text{Propionate}^- + 2H_2O \rightarrow \text{Acetate}^- + CO_2 + 3H_2$	+76,

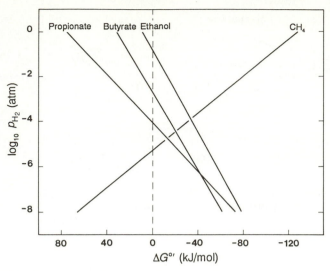

Fig. 2.24 Free energy change associated with the oxidation of propionate, butyrate and ethanol, and of methanogenesis, as a function of the partial pressure of H_2 (adapted from McInerney 1986).

and, as shown in Fig. 2.24, they do not become exergonic until the pH_2 is reduced below approximately 2×10^{-3} atm. (butyrate oxidation) or 9×10^{-5} atm. (propionate oxidation). The usefulness of having a methanogen as the syntrophic partner is also illustrated in Fig. 2.24, because H_2/CO_2 methanogenesis is still possible at H_2-pressures as low as about 10^{-5} atm.

Other types of organisms can also be syntrophic H_2-scavengers, e.g. H_2/CO_2 acetogens (homoacetogens) or sulphate reducers:

$$\Delta G^{0\prime} \text{ (kJ/mol)}$$

$$4H_2 + 2CO_2 \rightarrow \text{Acetate}^- + H^+ + 2H_2O \qquad -95$$
(Homoacetogen)
$$4H_2 + CO_2 \rightarrow CH_4 + 2H_2O \qquad -131$$
(Methanogen)
$$4H_2 + SO_4^{2-} + H^+ \rightarrow HS^- + 4H_2O \qquad -151$$
(Sulphate reducer),

but the lower free energy yield of the homoacetogen reaction obviously makes these organisms less suitable as syntrophic partners: the equilibrium reaction leaves very little, if any free energy to support growth of the other partner. As the syntrophic oxidation of volatile fatty acids proceeds *only* when the H_2-partial pressure is very low, the simultaneous presence of H_2-consumers and of inter-species H_2-transfer is obviously of fundamental importance.

2.3.3 *Methanogenic acetate oxidation*

Syntrophy is necessary for producing the relatively oxidized acetate from volatile fatty acids in freshwater environments and sewage digestors (i.e. environments in which sulphate is usually depleted) and, as we have already seen (Section 2.1), within communities of fermenters, a variety of fermentation pathways are funnelled towards the production of acetate. Thereafter, it is broken down principally by two groups of organisms — the acetoclastic methanogens, or if sufficient sulphate is available, by the sulphate reducers. There is however another method by which it can be dealt with — a methanogenic process which requires a non-methanogenic syntrophic partner; syntrophic acetate oxidation. The production of methane from the splitting of acetate was discussed in 2.2. The process is carried out by certain specialized methanogens (e.g. *Methanosarcina*) and it is, in principle, very similar to a fermentation — the methyl group of acetate ends up in CH_4, and the carboxyl group ends up as CO_2. An alternative process, also involving a methanogen, was originally proposed by Barker (1936). The first step was the oxidation of acetate to H_2 and CO_2, and the second, the reduction of CO_2 to CH_4 by H_2. It was originally believed that a single organism was responsible, until Zinder and Koch (1984) showed that two different organisms were involved. One of these, a non-methanogen, oxidized the acetate to CO_2 and H_2:

$$CH_3COO^- + 4H_2O \rightarrow 2HCO_3^- + 4H_2 + H^+ \qquad (\Delta G^{0\prime} = 105 \text{ kJ/reaction}),$$

whereas the partner, a methanogen, used the H_2 to reduce CO_2 to CH_4:

$$4H_2 + HCO_3^- + H^+ \rightarrow CH_4 + 3H_2O \qquad (\Delta G^{0\prime} = -136 \text{ kJ/reaction}),$$

giving an overall reaction for the process:

$$CH_3COO^- + H_2O \rightarrow CH_4 + HCO_3^- \qquad (\Delta G^{0\prime} = -31 \text{ kJ/mol},$$
$$\text{or about 16 kJ per partial reaction).}$$

Aside from the remarkable feat displayed by these two syntrophic partners living jointly from such a small energy yield, the process is also remarkable because of the identity of the non-methanogenic partner. It has been grown in pure culture and shown to be a homoacetogen (Lee and Zinder 1988). Homoacetogens typically depend on the H_2-dependent reduction of CO_2 to acetate, but this syntrophic organism exploits the low p_{H_2} to drive the reaction in the opposite direction. Thus the homoacetogen makes ATP whether it is engaged in either acetate formation or acetate degradation; the direction of the reaction presumably depending on the prevailing p_{H_2}. The relative importance of methanogenic acetate oxidation over acetoclastic methanogenesis in nature is unknown although in sulphate-depleted marine sediments there is some evidence (see Zinder and Koch 1984) that acetoclastic methanogenesis is not the only mode of acetate catabolism.

2.3.4 *Flocs and clusters*

The efficiency of uptake of H_2 or any other metabolite by a syntrophic partner will be a function of the concentration gradient of the metabolite and thus the distance over which it has to diffuse; so there is some advantage in the partners being positioned as closely as possible to each other. A variety of bacteria produce mucus, slime and other extracellular material. The reasons for doing this are rarely obvious, but in some cases it is apparent that the material serves to hold together syntrophic partners.

Tatton *et al.* (1989) established a continuous culture which was a stable association of three types of bacteria. In sulphate-free medium, they completely degraded ethanol to CH_4 and CO_2. *Desulfovibrio* fermented ethanol to acetate and H_2, so long as a *Methanobacterium* was present to remove the H_2; and a *Methanosarcina* cleaved the acetate to CH_4 and CO_2 (Fig. 2.25). The removal of acetate by *Methanosarcina* disposed of the need to introduce pH control and all three organisms could be grown in steady state up to a dilution rate of 0.8 day^{-1}, which was probably close to the maximum growth rate of the *Methanosarcina*. In this triculture, the *Methanosarcina* provided the extracellular fibrillar material which formed a matrix, trapping the *Desulfovibrio* and *Methanobacterium*.

A wealth of experimental evidence has been provided for the role of microbial aggregation in interspecies H_2-transfer. Conrad *et al.* (1985)

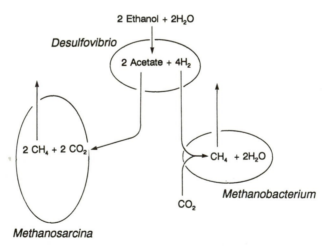

Fig. 2.25 Degradation of ethanol to CH_4 and CO_2 by a syntrophic consortium of three types of anaerobic bacteria in sulphate-free medium. Growth of the consortium achieved steady state, while supporting interspecies H_2-transfer and benefitting from pH control by the aceticlastic methanogens (based on information in Tatton *et al.* 1989).

Table 2.4. Free energy change (ΔG) calculated for H_2-consuming and H_2-producing reactions at *in situ* conditions in the anoxic sediment of Knaack Lake, a meromictic lake in Wisconsin, USA. The 'permissive p_{H_2}' is the partial pressure of H_2 at which the reaction concerned becomes exergonic (data obtained from Conrad *et al.* 1986)

	ΔG (kJ/mol H_2)	Permissive p_{H_2} (Pa)
H_2-consumption	-8.7	>0.11
Methanogenesis		
H_2-production		
Butyrate degradation	$+2.2$	<1.83
Propionate degradation	$+4.7$	<0.62
		In situ $p_{H_2}=4.2$

showed that in sewage sludge and in lake sediment, only about 5% of the H_2/CO_2-methanogenesis was driven by the bulk dissolved pool of H_2; the implication being that most of the methanogenesis depended on H_2 produced within flocs. Conrad *et al.* (1986) then proceeded to measure the *in situ* H_2-pressure in methanogenic lake sediments and various other sulphate-depleted anoxic habitats. They calculated the feasibility of fatty acid oxidation at the *in situ* p_{H_2}, and discovered that propionate and butyrate oxidation would be endergonic processes (Table 2.4). Nevertheless, these processes, and methanogenesis, were being performed within the sediment. This forced the conclusion that the *in situ* (i.e. the 'bulk') p_{H_2} was not the relevant parameter: fatty acid oxidation and methanogenesis were being carried out under conditions which were independent of the bulk conditions — they were being carried out within microbial clusters. Finally, convincing evidence supporting the importance of microbial cluster integrity was provided by Thiele *et al.* (1988). They isolated and washed the microbial flocs obtained from an anaerobic chemostat. These flocs had an average diameter of about 100 μm; they contained a variety of organisms, principally a *Desulfovibrio* species and a methanogen (*Methanobacterium* sp.), which together carried out the syntrophic conversion of ethanol to CH_4. For comparison, the conversion effected by the same organisms in the free bacterial fraction was also determined. The flocs produced methane at a higher rate and they produced only a low, stable level of H_2 in the medium (Fig. 2.26). On the other hand, the free bacterial community showed a high, transient production of H_2 and a lower rate of methanogenesis. When the integrity of the flocs was physically disrupted to produce smaller flocs with diameters of 10 to 20 μm, H_2-evolution increased, indicating diminished interspecies H_2-transfer associated with the loss of the structural integrity of the large flocs.

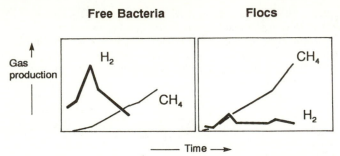

Fig. 2.26 Gas production by flocs and free bacteria engaged in anaerobic mineralization. The principal organisms involved were syntrophic partners — an ethanol-oxidizing *Desulfovibrio*, and a methanogenic *Methanobacterium* species. Results were obtained over a period of 60 hours (adapted from Thiele *et al*. 1988).

The most efficient syntrophic consortia will undoubtedly be those which maintain closely juxtaposed partners. And if there are large numbers of each partner in the consortium, they will, to sustain maximum efficiency, adopt whichever mosaic arrangement permits maximum possible mixing of both partners. The potential problem with this is that as each partner grows and divides, it may produce its own, clonal culture, which eventually becomes spatially segregated from its syntrophic partner. Progressive isolation of syntrophic partners will lead to a progressive reduction in the efficiency of the syntrophic metabolism. This has prompted the suggestion (Schink 1992*a*) that periodic physical disturbance may be necessary to re-mix the populations of syntrophic partners. But whether or not natural physical processes are capable of mixing individual bacterial clumps, each of which will, in all likelihood be attached to a substrate, is debatable. Of course, the problem would not exist if the different syntrophic partners were actually connected to each other, e.g. by filamentous material as described above (Tatton *et al*. 1989).

The phenomenon of intracellular syntrophy will be considered in detail later (Section 3.4), but a brief diversion is relevant here. The mosaic arrangements of syntrophic partners postulated by Thiele *et al*. (1988) and Schink (1992*a*) have actually been shown to exist within anaerobic ciliated protozoa. Figure 2.27 shows a hypothetical natural syntrophic consortium of bacteria, set alongside a transmission electron micrograph of part of the ciliated protozoon *Cyclidium porcatum*. The H_2-producing organelles in the ciliate are totally surrounded by H_2-consuming methanogens, and there is little doubt that an inter-'species' H_2 transfer (i.e. hydrogenosomes → methanogens) is operating in this, as it does in other anaerobic ciliates (see Section 3.4). But in none of these ciliates is there any evidence of progressive segregation of the syntrophic partners. It is true that

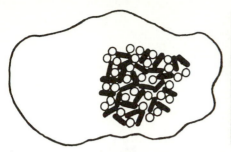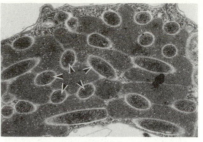

Fig. 2.27 *Left*: the close juxtaposition proposed for H_2-evolving (clear spheres) and H_2-consuming (black rods) syntrophic partners living in an anaerobic floc (outline). In the methanogenic floc shown, relatively little H_2 will escape to the bulk dissolved pool (adapted from Schink 1992*a*). *Right*: a transmission electron micrograph showing part of the anaerobic ciliated protozoon *Cyclidium porcatum* and the mosaic arrangment of H_2-producing organelles (hydrogenosomes) and methanogenic bacteria. Arrows indicate the direction of flow of H_2 from the former to the latter.

in some cases the two partners may be physically attached to each other, but it is also known that the two partners grow at the same rate. Even if there is no physical attachment between the partners in syntrophic consortia, if they are completely dependent on each other they must have similar growth rates, as shown in the last part of this section. This in turn must reduce the tendency of one partner to physically dominate a patch. In addition it is conceivable that the partners have evolved lattice geometries which minimize eventual segregation.

2.3.5 *H_2 or formate?*

Many of the methanogens and sulphate reducers which act as H_2-consuming syntrophic partners are also capable of oxidizing formate; and, as many of the H_2-evolving partners can also release reducing equivalents as formate, the suggestion that the latter could be the electron carrier mediating syntrophy has repeatedly been explored (e.g. Bryant *et al.* 1967; Thiele *et al.* 1988; Boone *et al.* 1989; Schink 1992*a*). The debate concerning whether H_2 or formate is the sole or principal means of electron transfer is difficult to resolve: there are serious technical problems in obtaining substantive experimental evidence; they have virtually identical redox potentials (see Fig. 2.1), the stoichiometries of the overall reactions are identical, and each is usually present at extremely low concentrations. It is also likely that they both reduce the same electron carriers, e.g. the coenzyme F_{420} of methanogens (Fig. 2.28). In some cases, formate cannot be oxidized by the methanogen, so H_2 must be the carrier; and the presence of a hydrogenation catalyst ($Pd-BaSO_4$) can facilitate butyrate oxidation when the methanogenic

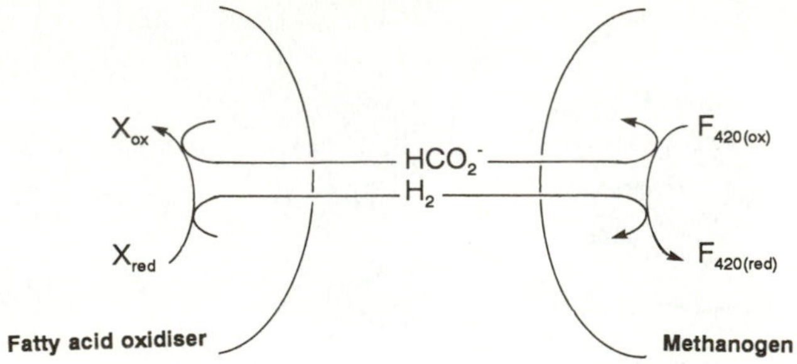

Fig. 2.28 Interspecies electron transfer between a hypothetical fatty acid oxidizer and a methanogen, utilizing either H_2 or formate (HCO_2^-). The reaction in the donor could be the re-oxidation of NADH, followed by reduction of the coenzyme F_{420} in the methanogen. Both mechanisms of electron transfer may operate simultaneously.

partner is inactivated (Kasper *et al.* 1987). The sulphate reducer *Desulfovibrio baarsii* cannot use H_2, so formate is the likely means of transfer (Zindel *et al.* 1988). At least one syntrophic amino acid fermenter (*Eubacterium acidaminophilum*) is known (Zindel *et al.* 1988) which will grow in co-culture with syntrophic partners capable of only H_2-consumption or only formate-consumption. It is also possible that syntrophy is facilitated by the transfer of various other metabolites. For example, elevated concentrations of acetate are known to impede its fermentation to acetone unless a syntrophic acetoclastic methanogen is also present (see Schink 1992*a*).

2.3.6 *Origin of syntrophy*

The syntrophic bacteria which carry out the oxidation of volatile fatty acids in methanogenic ecosystems are currently considered (McInerney 1992) to belong to four genera (*Syntrophobacter*, *Syntrophus*, *Syntrophospora* and *Syntrophomonas*), although it is likely that many more are waiting to be discovered and described. Most are very specialized in what they can do metabolically. Some can use only a single compound as their energy source; most do not degrade carbohydrates or proteins. It is widely believed that none are capable of anaerobic respiration, neither with organic nor inorganic electron acceptors (but, see below). Their natural substrates (volatile fatty acids) are already very reduced, so it is virtually impossible for them to produce oxidized intermediates that might serve as as electron acceptor. Thus, the production of H_2, or the reduction of CO_2 to formate (HCOOH) are probably the only mechanisms they have for disposing of reducing power. They produce acetate from even-numbered fatty acids (e.g. C_4, C_6), and acetate and propionate from odd-numbered acids. They are most important in aquatic habitats which are depleted in sulphate, e.g. freshwater sediments and sewage digestors. McInerney (1992) enumerated syntrophic butyrate-degrading bacteria in these habitats as 1000-fold greater than butyrate-degrading sulphate reducers. In marine environments, VFAs can be degraded directly by sulphate reducers (see Section 2.2) without the need for inter-species H_2-transfer. In the rumen, syntrophic bacteria can be found in relatively low numbers, but they grow too slowly to be maintained in this system. This, of course, is of some significance to the ruminant which depends on the VFAs for its supply of carbon and energy.

The evolutionary origin of such bacteria which are incapable of growth in the absence of a syntrophic partner is of considerable interest. It is likely that the first steps in solving the mystery have been taken with the recent determination of the base sequence of the 16S rRNA gene in *Syntrophobacter wolinii* (Harmsen *et al.* 1993). Comparison of this sequence with that for other bacteria indicates that it is most closely related to the sulphate-reducing bacteria, and that it shares a common ancestor with species of *Desulfomonile* and *Desulfoarculum*. In addition, it has been shown that *S. wolinii* can probably oxidize propionate to acetate when sulphate is available as an electron acceptor (see Harmsen *et al.* 1993). In other words, the organism is basically a sulphate reducer whose ancestors were sulphate reducers. In the absence of sulphate, it becomes a syntroph, still capable of oxidizing propionate to acetate, but only if it has a partner (even another sulphate reducer, e.g. *Desulfovibrio*) to consume the H_2 which is produced. *Syntrophobacter wolinii* is not closely related to the syntrophic butyrate oxidizers *Syntrophomonas wolfei* and *Syntrophospora bryantii*, implying that syntrophic fatty acid oxidation probably evolved independently on several occasions.

There is another, curious example of a symbiotic consortium which is of some relevance here. It involves two types of bacteria, and it is referred to as '*Chlorochromatium aggregatum*'. This consists of cells of a green photosynthetic bacterium resembling *Chlorobium limicola* which surround a colourless, polarly flagellated rod-shaped bacterium (Croome and Tyler 1984). The phototroph has been cultured: the rod-shaped bacterium has not. It has been suggested (Pfennig 1980) that it is a sulphur reducer. A similar consortium between a sulphur respiring-bacterium and a green sulphur bacterium is known as '*Pelochromatium*'.

One interesting aspect of the behaviour of the consortium is that the bulk of the planktonic population appears to move from the anoxic to the oxic zone, and *vice versa*. (Croome and Tyler 1984). This is unusual behaviour for prokaryotes living in a meromictic lake and it presumably reflects a rather dynamic symbiotic interaction between the two types of bacterium. The behavioural integration between the partners of the *Chlorochromatium* complex is also surprisingly sophisticated. The central rod provides the motility for the consortium, and this is integrated with the photosensitivity which resides in the phototroph. Nothing is known concerning the mechanism by which information about the underwater light climate is transferred between the two partners.

The functional or ecological significance of the syntrophic relationship between a sulphur-respiring heterotroph and a sulphide-oxidizing phototroph is that it allows for rapid internal cycling of sulphur within the aggregate in spite of a low ambient sulphur concentration. Athough on the borderline of the strict definition of syntrophy, a somewhat similar interaction between sulphate reducers and aerobic sulphide oxidizers plays an important role in the superficial layers of lake sediments which, because of rapid recycling, can sustain high rates of sulphate reduction in spite of the low ambient sulphate concentration in lake water (Section 4.1).

So far we have been largely concerned with obligate syntrophy, of the limited metabolic repertoire of syntrophs, and of their complete dependence on H_2/formate transfer. If we adopt a broader interpretation of 'syntrophy' however, we find many examples where two or more organisms grow together, both benefiting from H_2/formate transfer, where the syntrophy results in a lower production of electron sink products, but where both organisms are also capable of an independent existence. We find these examples in organisms which have a broader repertoire of fermentation pathways, and in particular, within those which carry out mixed acid fermentations (see Section 2.1). Although most of these organisms are prokaryotes, there are also some notable examples within the eukaryotes, especially the anaerobic chytridiomycete fungi (Bauchop 1989) and the anaerobic ciliated protozoa (Hillman *et al.* 1988), both of which live by fermenting cellulose and sugars in the rumen.

Taking the example of the chytridiomycete *Neocallimastix frontalis*, when cultured free of extracellular methanogens it produces acetate, lactate, ethanol, CO_2, H_2 and formate (Table 2.5). In co-culture with methanogens, H_2 is undetectable, acetate becomes the main product, and the production of lactate, ethanol and formate all decrease significantly. Thus H_2/formate consumption by the methanogens causes a shift in the flow of electrons away from electron sink products, and towards methane. The chytrid may also benefit from its increased energetic efficiency: the specific activities of extracellular hydrolytic enzymes increase 2–10-fold (Joblin and Williams 1991) in the presence of extracellular methanogens.

Similar results have been obtained with anaerobic ciliated protozoa (which also ferment, although with a narrower range of end-products; Hillman *et al.* 1988). Our discussion of syntrophy in the broadest sense, and in particular of how it impinges on the lives of anaerobic eukaryotes, continues in Chapter 3.4.

Table 2.5. Products of cellulose fermentation (mol/100 mol hexose units) by the anaerobic chytridiomycete fungus *Neocallimastix frontalis*, in the presence and absence of rumen methanogens (data taken from Bauchop and Mountfort 1981)

Product	*Neocallimastix* alone	*Neocallimastix* + Rumen methanogens
Acetate	73	135
Lactate	67	3
Ethanol	37	19
Methane	0	59
CO_2	38	89
H_2	35	< 0.05
Formate	83	1

2.3.7 *Energy sharing and equalization of growth rate constants of syntrophic partners*

Among the different types of interactions between pairs of species (other types of mutualism, prey–predator relationships, competition) syntrophy is peculiar in that it forces the partners to grow at the same per capita rate, and that on a per capita basis, the two species share equally the free energy of the overall process. The idea that growth rates tend to become equal has been pointed put by Powell (1984) and Archer and Powell (1985), and Schink (1992*a*) has claimed that energy is shared equally. Box 2.1 gives a formal and more general proof of these ideas.

It is shown that (in the case of exponential growth) the syntrophic partners will tend to increase at the same per capita rate and thus the individuals will share the energy of the overall bioenergetic process of the syntrophic consortium. The difference in energy yields between the bioenergetic processes of the two species is reflected in the population size ratio. This latter result is also suggested in a general way in Fig. 2.24 which suggests that the hydrogen tension will equilibrate at a level which yields equally much energy for each of the partners. The results apply irrespective of detailed assumptions about the functional forms of the growth responses to substrates and to metabolites as long as eq. [4] in Box 2.1 applies. The result also applies if species 2 not only removes a self-limiting metabolite of species 1, but also provides a substrate (e.g. recycling of sulphur, or if the phototrophic partner excretes organic substrates used by the heterotrophic partner). The model demonstrates in a general way that consortia such as *Chlorochromatium* are stable.

Box 2.1   Coupling of per capita growth rates of pairs of syntrophic species

Assume two species with the population sizes/biomasses x_1 and x_2. It is further assumed that the resources required by species 1 are unlimited, that this species produces metabolite M which limits its own growth, and at the same time, M is the resource of species 2. The growth of 1 and 2 is given by:

$$\mathrm{d}x_1/\mathrm{d}t = x_1\mu_1 f(M) \qquad [1]$$

$$\mathrm{d}x_2/\mathrm{d}t = x_2\mu_2 g(M) \qquad [2]$$

and
$$\mathrm{d}M/\mathrm{d}t = \mu_1 x_1 f/Y_1 - \mu_2 x_2 g/Y_2 \qquad [3]$$

where μ represents maximum growth rates, Y represents the growth yields (amount of x produced per unit M produced or consumed) and f and g are functions of M so that:

$$\partial f/\partial M < 0, \text{ and } \partial g/\partial M > 0, \ 0 < f, g > 1 \qquad [4].$$

The ratio of per capita growth rates is (from [1] and [2]) given by:

$$x_2\mathrm{d}x_1/x_1\mathrm{d}x_2 = \mu_1 f/\mu_2 g \qquad [5].$$

From eqs [1]–[4] it can be seen that if the concentration of M is increasing, the growth rate of x_1 will decrease and that of x_2 will increase until $\mathrm{d}M/\mathrm{d}t = 0$. Conversely, when $\mathrm{d}M/\mathrm{d}t < 0$, $\mathrm{d}x_1/\mathrm{d}t$ will increase and $\mathrm{d}x_2/\mathrm{d}t$ will decrease until $\mathrm{d}M/\mathrm{d}t = 0$. When M is constant, f and g are constant and so the ratio between [1] and [2] also becomes constant, and $\mu_1 f = \mu_2 g$ (from [5]). The per capita growth rates thus tend to equalize over time and the ratio between the population sizes x_1/x_2 approaches Y_1/Y_2 (from [3]).

In batch cultures and in chemostats, the per capita growth rates eventually become zero/negative or identical to the dilution rate, respectively. However, it can be shown that during the transient period (before equilibrium is reached) the per capita growth rates of syntrophic partners also tend to become equal.

2.4   The evolution of prokaryote energy metabolism

We have already discussed the evolution of some catabolic pathways (e.g. the origin of the citric acid cycle) in some detail in the two previous sections.

Here we consider how (or whether) current understanding of prokaryote genealogy yields insight into the earliest evolution of metabolic pathways, microbial communities and processes. The eukaryotes are considered separately, from a similar point of view, in Section 3.1 and 3.2.

The earliest attempts to classify bacteria were largely based on morphological properties in analogy with the classification of plants and animals. A few bacterial groups which were established in this way still hold as phylogenetic entities (e.g. the spirochaetes) and morphological features remain important for the identification of species or higher taxa within some groups of prokaryotes. However, it was soon recognized that bacterial morphology (as revealed by the microscope) is in most cases too simple and crude to serve as the basis for classification or identification. A variety of other phenotypic traits were therefore used and among them, properties of metabolism were prominent. Traits which are considered important include aerobic versus anaerobic metabolism and whether the organisms are phototrophic, chemolithotrophic or organotrophic, the types of substrates which the organisms utilize, and types of metabolites which are excreted. Such traits are still important for determinative bacteriology and they will remain the basis for an operational species concept in prokaryotes.

Throughout most of this century it was also widely believed that bacterial phylogeny would eventually be understood in terms of the evolution of energy metabolism. This was in part reflected in classification schemes, so that some functionally defined organisms such as the phototrophic bacteria were grouped together. Attempts were also made to understand how different catabolic and anabolic pathways might have evolved. The discussion was often flavoured with the idea originally proposed by Oparin (1953) (Section 1.2), that the first organisms made their living by fermenting carbohydrates in the 'primordial soup'. Thus clostridial fermentation was considered to be the most primitive type of energy metabolism while phototrophic and respiratory metabolism were believed to have evolved later.

Since gene sequencing was introduced in the 1970s, our view of prokaryote evolution has changed. At first, these studies had a somewhat limited scope and were based on the amino acid sequences of molecules such as cytochrome *c*. More recently, the sequencing of conservative genes, and especially of 16S rRNA, has revolutionized our understanding of bacterial phylogeny. Detailed discussions of the history of bacterial classification and phylogeny as well as rRNA geneaology are found in Stackebrandt (1992) and Woese (1987, 1992).

There are several technical difficulties involved in molecular geneaology (for references see Woese 1992). A more fundamental problem is whether rRNA-sequences reveal the phylogeny of the organisms rather than the phylogeny of the rRNA gene itself. Bacteria could be a bag of promiscuous genes which are transferred horizontally. If this was the case, only genes, but not bacteria would have a phylogeny. The significance of rRNA-sequencing would then be much more limited and the following discussion would be very different, or meaningless.

The problem has been considered in some detail by Woese (1987). There is no doubt that plasmid genes spread horizontally and that this may also be the case for chromosomal genes. A well-known case is the gene for nitrogen fixation which occurs in a plasmid. The trait is found in a variety of entirely unrelated bacteria, but the phylogeny of the gene will not reflect the phylogeny of its bearers. However, in other cases a close correlation between the phylogeny of different genes (e.g. between cytochrome *c* and rRNA) has been found. The general conclusion of Woese (1987) is that bacteria do have a phylogeny and that only a small fraction of genes spread horizontally. In what follows, we will implicitly assume that bacteria have a phylogeny and that cases with an apparent absence of correlation between (rRNA) geneaology and phenotypic traits must be explained by means other than horizontal gene transfer. However, the exact fraction of genes which are transferred horizontally is unknown and the problem remains a caveat when considering prokaryote evolution.

Molecular phylogeny has first of all shown that all living organisms are related and that all extant life can be classified into three main groups: the eubacteria, the archaebacteria and the eukaryotes (Fig. 2.29). The two prokaryote groups are now often referred to as Bacteria and Archaea, respectively in order to emphasize that they are mutually no more closely related than is either group to the eukaryotes. All three groups have had a very long independent evolutionary history and they may have diverged shortly after the origin of the first cells.

The second general conclusion which can be drawn from rRNA genealogy is that the phylogenetic groupings frequently do not correlate with pheno-

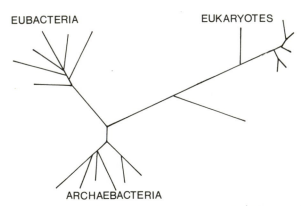

Fig. 2.29 The universal phylogenetic tree (modified after Woese 1987).

typic traits such as energy metabolism or at least the correlations are different from what had once been expected. Many major groups include different forms which are fermenters, phototrophs, chemolithotrophs or respirers and conversely, groups which were previously considered monophyletic on the basis of physiological properties (e.g. phototrophs or sulphate-sulphur-reducing forms) now seem to contain forms which are only remotely related. The consequence of this is that many types of energy metabolism (e.g. respiration) must have evolved independently within several lineages, from a single or a few basic kinds of ancestral energy metabolism. Evidence suggests that the ancestral eubacterium was an anoxygenic phototroph and this may also apply to the common ancestor of the eubacteria and the archaebacteria. The new phylogenetic tree also suggests that the ancestral forms were anaerobes and that aerobes have shallower roots, and they have arisen many times within different lineages. Their apparently more recent origin probably reflects the appearance of atmospheric oxygen about 3.5 billion years ago (Stackebrandt 1992). The eukaryotes were also anaerobes prior to their acquisition of prokaryote endosymbionts which eventually became mitochondria; in the absence of these organelles only fermentative energy metabolism occurs in extant eukaryotes (see Sections 3.1 and 3.2).

The apparent lack of correlation between phenotypic traits and geneaology is reflected in the phylogenetic tree for the eubacteria (Fig. 2.30). It is seen that phototrophic forms appear within five major groups with deep roots. The Gram-positive bacteria include only one phototrophic genus (*Heliobacterium*). The purple bacteria also include a majority of non-phototrophic forms while the three other groups (green sulphur bacteria, green non-sulphur bacteria and cyanobacteria) include only phototrophs. Apart

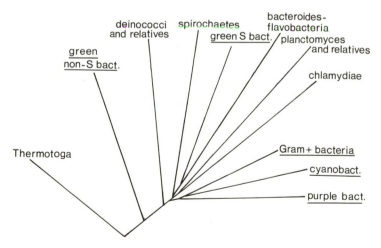

Fig. 2.30   The phylogeny of the eubacteria (redrawn after Woese 1987). Groups containing phototrophs are underlined.

from the oxygenic cyanobacteria and the microaerophilic purple phototrophs, the phototrophic bacteria are all anaerobes. It is difficult to imagine
the independent origin of chlorophylls and reaction centres within different
lineages whereas it is easier to see how some types of chemotrophic energy
metabolism could have evolved from phototrophy (see below). The general
conclusion is therefore that the ancestral eubacterium was an anoxygenic
and anaerobic phototroph (Woese 1987).

A more detailed look at one of the major eubacterial groups, the purple
bacteria (also referred to as the proteobacteria), emphasizes the central role
of phototrophy in eubacterial evolution. The group is referred to as 'purple
bacteria' because the ancestral form must have been a purple (phototrophic)
bacterium, but relatively few of the extant members are phototrophs. In
fact, many of the well-known Gram-negative bacteria (various chemotrophic
aerobic or faculative anaerobes including enterobacteria and several human
pathogens, as well as many sulphate reducers and chemolithotrophs) belong
to the purple bacteria (Fig. 2.31). These bacteria form four major natural
groups (referred to as the α, β, γ and δ groups). The α and β groups include
the purple non-sulphur bacteria as well as a variety of organotrophic and
chemolithotrophic bacteria. The ancestor of eukaryote mitochondria
also belonged to the α-group. The γ-group includes the purple sulphur
bacteria, some sulphide and sulphur-oxidizing chemolithotrophs and
some organotrophic respirers. The δ-group does not include phototrophic
forms, but several sulphate reducers as well as aerobic organotrophs. The

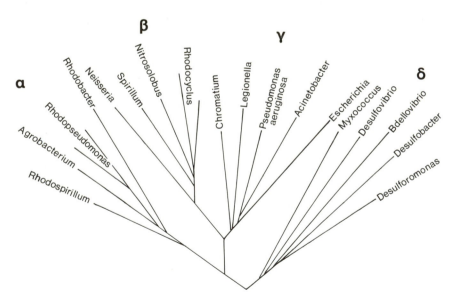

Fig. 2.31 The phylogeny of the purple bacteria (redrawn after Woese 1987).

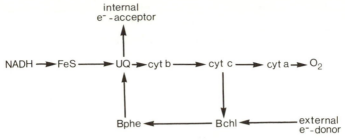

Fig. 2.32 Electron transport in the cell membrane of a purple non-sulphur bacterium (FeS, iron-sulphur protein; UQ, ubiquinone; cyt, cytochrome; Bchl, bacteriochlorophyll; Bphe, bacteriopheophytin) (based on an idea in Whatley 1981).

implication is again that phototrophs have independently given rise to a variety of heterotrophic or chemolithotrophic forms within each of the four groups.

That the conversion from phototrophy to respiration may be mechanistically simple is suggested in Fig. 2.32, which shows electron flow in a purple non-sulphur bacterium. The organism can perform light-driven cyclic phosphorylation, thus producing a proton gradient across the cell membrane, the potential energy of which can be conserved as ATP. The photosynthetic apparatus can also accept electrons from external electron donors (e.g. H_2) which are then diverted to biosynthetic purposes. The electron transport system can also receive electrons deriving from a substrate (via NADH) and pass them on to an external electron acceptor (e.g. O_2), thus generating a H^+-gradient; a process by which the organism generates energy in the dark. If such an organism lost its capability to produce photosynthetic pigments it would be transformed into a chemotrophic aerobe. In a similar way it is possible to imagine how purple sulphur bacteria, for example, developed into chemolithotrophic sulphide oxidizers, or how changes in cytochromes allowed for the utilization of different terminal electron acceptors (fumarate and sulphate, and, later in evolution, oxygen and nitrate).

The idea that anoxygenic phototrophy must be a very ancient type of energy metabolism from which all other types have evolved is not entirely new, but Olson and Pierson (1987) and Pierson and Olson (1989) have more recently presented arguments in favour of the idea. Considering the alternative hypothesis (that fermentation represents the original type of energy metabolism) it can be noted that fermentation pathways such as glycolysis are not mechanistically simple. They involve many different enzymes and it is hard to imagine how they could have originated 'from scratch'. It also seems unlikely that sufficient amounts of carbohydrates were produced in the primordial 'soup' to sustain life based on carbohydrate fermentation (carbohydrates are not easy to produce in 'prebiotic life experiments'; see Section 1.2).

It seems more likely that when the production of high-energy phosphate bonds by photochemical processes in the environment could no longer supply the necessary energy for the earliest life, these photochemical processes became incorporated into the cell membrane. As pointed out by Pierson and Olson (1989) a functional photosynthetic reaction centre could be mechanistically simple, involving in principle only a membrane with a photo-excitable porphyrin molecule, an FeS-protein and a quinone. This could act as a photo-driven electron transport chain which would create a proton gradient (Fig. 2.33), but it would require both an external electron donor and an external electron acceptor. The addition of a cytochrome (the precursor of cytochrome *c*) would allow for cyclic phosphorylation and, after the development of the first chlorophyll (from the hypothetical porphyrin), the system would be transformed into a reaction centre, as found in extant phototrophic bacteria.

Fermentative pathways occur, of course, among organisms which are entirely dependent on fermentation and substrate phosphorylation for energy conservation, but they are also widely distributed among anaerobic and aerobic respirers, in which they generate and feed NADH into the electron transport system. As shown in Section 2.1 there is a considerable variety of fermentative pathways and many of these are found in quite unrelated organisms. The evolution of these pathways is still incompletely understood. It can be speculated that they were originally used for biosynthetic purposes and that they developed in the earliest phototrophs when various necessary molecules could no longer be assimilated directly from the environment. Such pathways could develop stepwise in response to the sequential depletion of different precursors in the environment. Only later

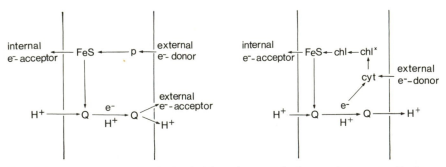

Fig. 2.33 *Left*: a hypothetical primitive photosynthetic reaction centre. *Right*: a cytochrome has been added which allows for cyclic photophosphorylation. (FeS, iron-sulphur protein; Q, quinone; p, protoporphyrin; chl, a chlorophyll; chl*, an excited chlorophyll molecule; cyt, cytochrome) (based on Pierson and Olson 1989).

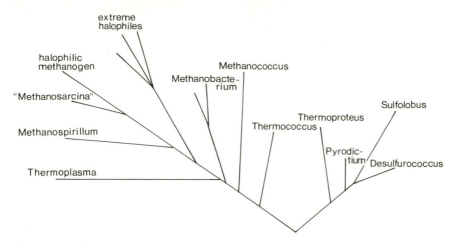

Fig. 2.34 The phylogeny of archaebacteria (redrawn from Woese 1987).

did they function in reverse, generating energy by substrate level phosphorylation. An example of this is the evolution of the citric acid cycle which occurs as a reductive cycle for biosynthesis in green sulphur bacteria and in some archaebacteria. Only later, it may be assumed, did the citric acid cycle function as a catabolic oxidative cycle in respiratory organisms, by running in reverse (Section 2.2). That metabolic pathways or cycles can be changed to serve other functions, and that this can take place over a relatively short evolutionary time span (much shorter than otherwise considered in this section), is seen in the case of some metazoa which depend on anaerobic metabolism (Section 3.5). In various nematodes and trematodes part of the citric acid cycle is used in reverse to supplement anaerobic energy generation.

Turning to the archaebacteria (Fig. 2.34), they are, as a group, seemingly less diverse than the eubacteria with respect to energy metabolism. Functionally, they include three main groups: the methanogens, the extreme halophiles and the extreme thermophiles. The extreme halophiles live in hyperhaline habitats and thrive even in saturated NaCl-solutions; the extreme thermophiles have growth optima in the range 70–100 °C and can be isolated from hot springs and similar habitats. The energy metabolism and unusual biochemical features of methanogens were discussed in detail in Section 2.3. The extreme halophiles are aerobic chemoheterotrophs. The energy metabolism of the extreme thermophiles is in a few cases fermentative, but in most species it is based on sulphur. Some are anaerobes which use elemental sulphur (or more oxidized S-compounds) as electron acceptors in respiration, with various organic compounds as substrates. Others are aerobes which oxidize organic compounds or reduced sulphur compounds.

The group exemplifies how the metabolic machinery has diversified to carry out what were once believed to be fundamentally different or opposing types of energy metabolism (anaerobes vs. aerobes, oxidizers vs. reducers of S-compounds). Phototrophy is not found among the archaebacteria except in the extreme halophiles. In these, it is not based on chlorophylls, but on a carotenoid (see Section 2.2), and it is believed to have evolved independently within this group (Woese 1992). The extreme halophiles do, however, provide an example of how a mechanistically simple, light driven proton pump can evolve.

It is now generally concluded that the phylogenetic tree of the archaebacteria shows that the ancestral archaebacterium was an anaerobic sulphur respirer and that the S-oxidizing thermophiles, the methanogens and the halophiles are derived from them. It is also believed that the ancestral archaebacterium was a thermophile (Woese 1987). This suggestion, together with the fact that an early branch of the eubacteria (*Thermotoga*) is also a thermophile, has inspired some authors to suggest that the Earth was a warm place during the early Archaean and that the origin of life took place in hot environments (Gottschal and Prins 1991). This, however, remains speculation. Geothermal areas in shallow seas and hot springs must have persisted throughout the history of the Earth and the ancestors of the extant extreme thermophiles may simply have adapted to this specialized niche early in the history of life.

If we consider the eubacteria and the archaebacteria together, it is clear that their common ancestor had already acquired the fundamental components of metabolism including an electron transport chain with FeS-proteins, quinones and cytochromes and (accepting that they originated as a part of a photosynthetic reaction centre) chlorophylls, as well as some of the basic biosynthetic pathways. Relatively smaller modifications of these components would have later allowed the use of different terminal electron acceptors (O_2 and NO_3^-) for respiration and the development of fermentative pathways in order to utilize a wider spectrum of organic substrates, or as the sole mechanism for energy generation. These developments took place independently within several lineages and probably at different times during evolution.

Returning to the questions posed in the beginning of this section, the phylogenetic scheme which has resulted from rRNA nucleotide sequencing does confirm our belief that life originated and diversified under anaerobic conditions. This early diversification included not only the branching which resulted in the origin of eubacteria, archaebacteria and eukaryotes, but probably also the origin of the deepest branches within each of these groups. We can also conclude that all the fundamental biochemistry of metabolism had already evolved in the common ancestor of the three major groups of life.

On the other hand, it is also evident that metabolic pathways have frequently been modified to adapt to quite different types of energy metabolism during evolution. The adaptation of electron transfer systems to use oxygen as a terminal electron acceptor has happened independently in many groups and conversely it is likely that many organisms have secondarily adapted to an anaerobic life style. It is possible that some apparently 'ancient' forms such as green non-sulphur bacteria and methanogens have remained phenotypically unchanged since well before the advent of cyanobacteria and atmospheric oxygen (> 3.5 billion years ago). But in general, we cannot take it for granted that the organisms on the 'ends of the deep branches' have remained phenotypically unchanged throughout the history of life. Nor can we assume that the types of prokaryotes we find in anaerobic habitats today all existed before the oxic era of the Earth. Many of the extant prokaryotic anaerobes may secondarily have adapted to anaerobic life or they may otherwise have changed phenotypically in substantial ways during more recent evolutionary time. It is now clear that most extant anaerobic eukaryotes are descended from aerobic ancestors (Sections 3.1 and 3.2). Among the prokaryotes this is known with certainty in only a few cases (e.g. the denitrifiers; see Sections 1.2 and 2.2), but it is likely to have happened repeatedly within many lineages.

As discussed in Section 1.2, the conditions of life changed radically after the advent of oxygenic photosynthesis, not only by allowing for aerobic respiration and for a vast increase in biological productivity; the advent of atmospheric oxygen also changed properties of anaerobic habitats and the conditions for anaerobic life (Sections 1.2 and 5.1). On the basis of current knowledge, considerations of the phenotypic properties of organisms which were important prior to the origin of oxygenic photosynthesis are still mainly speculation. In this particular area our recent enhanced understanding of prokaryote phylogeny has had limited impact.

3

Anaerobic eukaryotes: phagotrophy and food chains

3.1 The origin of eukaryotes

Endocytosis is a fundamental property of eukaryotes. While some analogies to predation and parasitism are found in the prokaryote world (e.g. the 'predatory' bacterium *Bdellovibrio*, see Guerrero *et al*. 1986) the original ecological niche of eukaryotes was probably one of predation, and the advent of eukaryotes introduced food chains into ecological systems.

This chapter will deal, first of all, with eukaryotic biota in contemporary anaerobic habitats and the biology and physiology of anaerobic eukaryotes. In the first two sections, however, we will briefly discuss the origin of eukaryotes and the phylogeny of anaerobic eukaryotes. The possibility that the latter can somehow illuminate the early evolution of eukaryotes will be examined. Regarding the prokaryotes, genealogical trees based on rRNA-sequences suggest that anaerobic forms have 'deep roots' whereas different aerobes seem to have evolved independently at a later time when the atmosphere became oxic (see Section 2.4). Thus, most major groups of prokaryotes had diverged prior to the advent of an oxygen-containing atmosphere. The conclusion of this and the following sections is, however, that while extant anaerobic eukaryotes may provide useful analogies to early eukaryotic evolution, most are in fact descended from aerobic ancestors and have secondarily adapted to life without oxygen. A few groups of anaerobic protozoa are relatively distant relatives of other eukaryotes, and their lack of mitochondria may be a primary characteristic. In some other respects, however, even these organisms have typical and complex eukaryotic features (eukaryotic flagella, microtubules, mitosis, nuclear envelope) and a convincing link between eukaryotes and prokaryotes remains to be found.

3.1.1 *The earliest evolution of eukaryotes*

Figure 3.1 shows the 'universal tree' as based on rRNA-sequences (only the eukaryotes are presented in any detail). The tree is unrooted in the absence of an outgroup. It is striking that the genetic distance between the three major groups of life (eubacteria, archaebacteria and eukaryotes) is so large and that the genetic divergence among the eukaryotes is relatively large compared to the two other groups. The general conclusion of this is that the eukaryotes have a deep root; that is, that they diverged very early from the two prokaryote groups. It seems likely that the last common ancestor must have been close to the first real cell; some authors have even suggested that this last common ancestor was a 'progenote' (prior to the establishment of the genome) or even a representative of the RNA-world (Forterre *et al.*

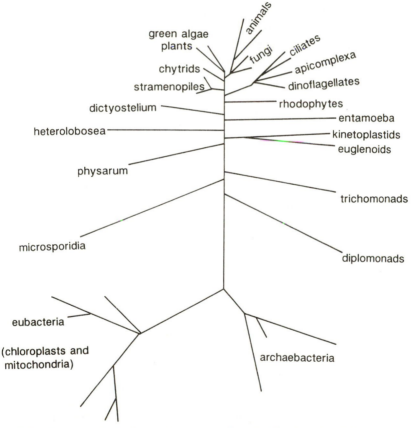

Fig. 3.1 A universal phylogenetic tree based on small sub-unit rRNA sequences. The figure was compiled from various sources cited in the text.

1993). However, as shown in Section 2.4 the common ancestor had already evolved most basic biochemical pathways.

In some respects the eukaryotes resemble the archaebacteria more than they resemble the eubacteria (e.g. Zillig 1987). The thermophilic archaebacteria (*Thermoproteus, Sulfolobus, Desulforococcus*) in particular have drawn attention in this respect. These organisms do not have a cell wall, but a flexible cell membrane with some sort of cytoskeleton (Searcy 1987; Searcy and Hixon 1991) and they share some other traits (ribosomal structure, introns) with the eukaryotes (Lake 1988), but they do not have a eukaryotic nucleus or a nuclear envelope. Lake (1988) has re-interpreted the rRNA data to make the thermophilic archaebacteria a sister group of the eukaryotes, creating the term 'karyotes' to designate this pair of groups. (It has become a questionable practice to create new names for various putative groupings every time a new, often interesting, but fanciful phylogenetic theory is published. The application of methods of molecular biology has made great strides in our understanding of the evolution of life, but so many facts are still open to interpretation and much more data are needed before any consensus can be reached. (And, by that time, the real purpose of a taxonomic nomenclature may have been undermined by a plethora of new names.) At any rate, one of the less attractive aspects of deriving the eukaryotes from the thermophilic archaebacteria is that most of the latter have an energy metabolism based on sulphur although some have fermentative metabolic pathways.

Various theories have been put forward to explain the origin of the eukaryotic nucleus, the nuclear envelope, mitosis, the cytoskeleton, etc. (e.g. Cavalier-Smith 1987*a*). Some of these may contain some truth, but they are generally not very convincing, being more the expression of topological imagination than testable biological theory.

Under all circumstances it is reasonable to assume that the ancestors of eukaryotes had diverged very early in the history of life and that by the time they acquired mitochondria, they were phagotrophic, anaerobic cells which depended on fermentation for energy generation (Müller 1992; see also Sections 3.2 and 3.4).

3.1.2 *The acquisition of organelles*

The origin of the characteristic cell organelles of eukaryotes (centriole/cilia, mitochondria, chloroplasts, peroxisomes) has often been considered to be more or less synonymous with the origin of eukaryotes. Regarding the chloroplasts and mitochondria, some speculative ideas on their origin as prokaryote endosymbionts which were suggested early in this century, were taken seriously by Margulis (1970, 1993). This proved to be very successful in terms of explaining the properties of these organelles (organelle DNA of bacterial type, sensitivity to certain antibiotics, similarity between organelle and eubacterial nucleotide sequences and a variety of other characteristics) and also in inspiring new research on these organelles and on intracellular symbionts. Today, the interpretation of chloroplasts as being derived from endosymbiotic cyanobacteria and of mitochondria as being derived from endosymbiotic oxygen-respiring bacteria is generally accepted (e.g. Delihas and Fox 1987).

Regarding chloroplasts, evidence suggests that some eukaryotes had acquired them by the mid Proterozoic and it is likely that the evolution of chloroplasts from 'endocyanelles' has happened independently several times. This is also supported by extant examples of endosymbiotic cyanobacteria and of intermediates between endosymbionts and organelles (Schenk *et al.* 1987).

On the basis of the smaller size of mitochondrial DNA and the higher degree of integration in the cell, mitochondria may represent a more ancient acquisition (Woese 1977); although it has also been argued that the different organelles were acquired at the same time (Cavalier-Smith 1987*b*). Mainly on the basis of morphology (the mitochondria of most protists have tubular cristae while those of animals, plants, fungi and a few protist groups have flattened cristae) it has been argued by Stewart and Mattox (1984) that mitochondria are of polyphyletic origin. All evidence suggests that the ancestor(s) of the mitochondrion belonged to the α-group of the purple bacteria. These bacteria include photosynthetic non-sulphur bacteria (which are microaerobic respirers in the dark) as well as aerobic heterotrophs. The α-group of the purple bacteria and mitochondria show close similarities with respect to the electron carrier chain, the amino acid-sequence of cytochrome *c* and rRNA-sequences (Dayhoff and Schwartz 1981; Delihas and Fox 1987; Yang *et al.* 1985). The outer and inner membrane of the mitochondrion is usually interpreted as the host vacuole and the symbiont cell membrane, respectively. Alternatively, the double membrane may represent the inner and outer membrane of a Gram-negative bacterium which was not contained in a host vacuole (Cavalier-Smith 1987*b*); this interpretation is supported by an extant example of an endosymbiotic purple bacterium, which lives freely in the host cytoplasm (see Section 3.4).

The ancestor of the mitochondrion has also been considered to be an aerobic respirer (such as *Paracoccus*) which established itself in a fermentative eukaryotic cell (Whatley 1981). Woese (1977) pointed out that this may represent a paradox, because neither the symbiont nor the host would gain from such an association in which the host was sensitive to oxygen and lived in anoxic habitats. Rather, he suggested the more comprehensive adaptive significance of a phototrophic organism living symbiotically inside a heterotroph. The ancestor would then be a purple non-sulphur bacterium (like *Rhodopseudomonas*) and the significance in terms of respiration would be secondary. This is supported by the extant analogy of a prokaryotic endosymbiont which confers the ability of oxygen respiration to an anaerobic, eukaryotic host: a ciliate with endosymbiotic purple non-sulphur bacteria (Fenchel and Bernard 1993*a,b*). However, there is no reason to exclude the possibility that eukaryotes had already acquired oxygen tolerance before they acquired mitochondria. In addition, the respiratory activity of endosymbionts may have protected the host from oxygen toxicity in microaerobic habitats.

It is not known when eukaryotes first harboured mitochondria. If cytochrome *c* is used as a molecular clock (calibrated by animal cytochrome *c* sequences) the divergence from purple bacteria may have happened 1.5–2 billion years ago, but such estimates are very uncertain and it could have happened much earlier. Today, practically all eukaryotes (including most anaerobes) have mitochondria or organelles derived from mitochondria and only a few groups are amitochondriate (see next section). As discussed in more detail in Section 5.3, the combination of aerobic respiration and phagotrophy is one of the most important events in the evolution of life: it allowed for the appearance of animals and for ecological systems with complex food chains.

For the sake of completeness it should be mentioned that attempts to explain the existence of some other eukaryotic organelles in terms of symbiosis with bacteria have been made. Thus, de Duve (1969) and Cavalier-Smith (1990) suggested that peroxisomes (which remove oxygen and peroxide in the cell in processes which are not coupled to energy conservation) are descended from a prokaryote endosymbiont which rendered the eukaryotes oxygen-tolerant before they acquired mitochondria. In the absence of substantial evidence (e.g. the presence of organelle DNA) this idea has not gained general acceptance. Margulis (1993 and papers cited therein) has suggested that eukaryotic flagella/cilia and the centriole derive from ectosymbiotic spirochaetes. This hypothesis has not attracted many followers; the evidence for an organelle genome is at best dubious and spirochaetes bear nothing more than superficial resemblance to cilia.

3.2 Phylogeny of anaerobic eukaryotes

During the last decade it has become increasingly clear that the traditional classification of the eukaryotes does not reflect well their evolutionary history. New morphological information, especially that obtained with transmission electron microscopy, has shown that protistan groups, like the flagellates and the amoebae, are polyphyletic and include a larger number of independent lineages which must have diverged very early during the evolution of the eukaryotes. Moreover, the allocation to botanists, zoologists and mycologists, respectively of green, motile/heterotrophic and fungi-like protists, has been shown to have no phylogenetic meaning and the chloroplast-containing forms do not form a monophyletic group (Corliss 1984; Patterson and Larsen 1991). Thus Corliss (1984) recognized 45 phyla of protists (eukaryotes which are not animals, higher plants or higher fungi).

This view has been vindicated by the recent rapid accumulation of data on small subunit rRNA-sequences (Schlegel 1991; Sogin 1991; Sogin *et al.* 1986; see also Fig. 3.1). Representatives from several important groups (e.g. foraminifera, radiolaria and many other representatives of the rhizopods, choanoflagellates, and rhizomastigids) remain to be sequenced. This is likely to happen within a few years and it will probably change the appearance of Fig. 3.1. It should also be mentioned that the construction of phylogenetic trees may be based on different algorithms which yield different results, and other factors also complicate the interpretation of the data (for a discussion on the methods and for references, see Schlegel 1991). But several important conclusions can be drawn. There is considerable genetic distance between the main groups of the eukaryotes, and organisms such as the flagellates encompass entirely unrelated groups (terms such as 'flagellates', 'slime moulds', 'amoebae' remain useful, however, to describe organizational levels or cell types). The animals, plants and fungi are relatively closely related to each other and each of these groups presents a relatively modest genetic divergence (comparable to that within the ciliates, for example, many of which have now been sequenced). Some groups of eukaryotes diverged very early in the evolution of life. Figure 3.1 also gives the impression that there have been special episodes during the history of the eukaryotes with bursts of adaptive radiation and the initiation of new lineages. The relationship between the rise of atmospheric p_{O_2} and the evolution of metazoa, and between the later colonization of land and the radiation of higher plants and fungi (see Sections 1.2 and 4.2) are such events. However, the fact that only few representatives of the more divergent eukaryotes have so far been sequenced may in some cases give a false impression of such special episodes during evolution.

In their quest for the most primitive eukaryote, it has not escaped the notice of several authors that at least two of the early lineages of the eukaryotes are amitochondriate. While we will show that anaerobicity has evolved secondarily in many cases among 'higher' unicellular eukaryotes (especially among the ciliates), some amitochondriate protists may actually be descended from, and resemble, forms which never acquired mitochondria. In some other cases, this interpretation is not very convincing.

3.2.1 The amitochondriate protists

Major groups of anaerobic protists or groups which include anaerobes as well as aerobes are listed in Table 3.1. At least three major groups of eukaryotes do not have mitochondria. Their diversity and morphology are reviewed in Mylnikov (1991) and Brugerolle (1991), respectively. In the cases where energy metabolism has been studied (e.g. the intestinal commensal *Giardia*) it is based on glycolysis and mixed acid fermentation (Müller 1988; see also Section 3.4). The rhizomastigids include the giant amoeba *Pelomyxa*, and *Mastigella*, *Mastigamoeba* (see Fig. 4.18), and *Phreatamoeba*. *Pelomyxa*, which occurs in lake sediments, has many short, non-motile flagella; the other three, much smaller forms, each have one long, motile anterior flagellum. They possess neither mitochondria nor a Golgi apparatus, nor microbodies. The rRNA-sequence is not yet available for any representative, but their morphology indicates that they may, like the diplomonads, represent a very ancient branch of the eukaryotes, and the absence of

Table 3.1. Protist groups with anaerobic representatives

	Mitochondria/mitochondria-like organelles[1]	Lifestyle[2]
1. All species anaerobic		
Rhizomastigidae	−	FL
Diplomonadida	−	FL/IC
Retortomonadida	−	FL/IC
Trichomonadida	+(H)	FL/IC
Oxymonadida	?	IC
Hypermastigida	+(H)	IC
Entamoeba	−	IC
Incertae sedis:		
Psalteriomonas[3]	+(H)	FL
Blastocystis	+(?)	IC
2. Some species anaerobic		
Heterolobosa[4]	+?	FL
Chrysomonadida	+?	FL
Kinetoplastida	+?	FL
Choanoflagellida	+?	FL
Chytridiomycetes[5]	+(H)	IC
Ciliophora	+(+/-H)	FL/IC

[1]H, hydrogenosomes. [2]FL, free-living; IC, intestinal commensals. [3]May belong to the Heterolobosa. [4]Representatives for this and the three following groups have been found growing under anaerobic conditions, but not studied in detail. [5]Chytrids occur as free-living forms, but anaerobic species are known only from the cow rumen.

mitochondria may be a primitive feature, as suggested by Cavalier-Smith (1991*a,b*). The diplomonads are common phagotrophic inhabitants of anaerobic habitats (e.g. *Hexamita, Trepomonas*; see Fig. 4.18) while others (*Giardia*) are intestinal commensals/parasites (e.g. in man). Molecular data suggest that they represent the earliest known branch of the eukaryotes. Their morphology seems more complex than that of the rhizomastigids (they have multiple flagella and the cells form peculiar mirror-symmetric or rotationally-symmetric doublets with two nuclei and two sets of flagella). But they do not have mitochondria, a Golgi apparatus, an endoplasmic reticulum or microbodies (Brugerolle 1991). Their rRNA has many segments which resemble archaebacterial sequences and the cytoskeleton is simple (Schlegel 1991). The retortamonads occur mainly as intestinal commensals, but the genus *Chilomastix* (see Fig. 4.18) also includes anaerobic, free-living representatives. Their morphology is more complex and in different respects they resemble both the trichomonads and the heterolobosans (Brugerolle 1991). No molecular data are yet available.

Two other groups of protists are less likely candidates as primarily amitochondriate eukaryotes. Members of the genus *Entamoeba* occur exclusively in the intestinal tract (e.g. in man where one species causes dysentery). They rely on mixed acid fermentation for energy generation (Müller 1988). Their position in the evolutionary tree (Fig. 3.1), however, suggests that they branched off after eukaryotes had acquired mitochondria (kinetoplasts, euglenoids, and the heterolobosans are all mitochondriate). The organism probably exemplifies the phenomenon of loss of mitochondria during evolution. Finally, the microsporidia seem to have branched off early in eukaryote evolution (Fig. 3.1). In many ways they are aberrant eukaryotes, they lack 5.8 S rRNA and almost all other characteristics of eukaryotic cells except for a nuclear envelope. However, they show little sequence similarity with prokaryotes (in contrast to the diplomonads) and they have probably undergone a rapid genetic evolution, which complicates the interpretation of their phylogenetic position (Schlegel 1991). Their characteristic features should probably be understood in terms of the fact that they are all intracellular parasites of other eukaryotes and they have therefore been able to dispense with and subsequently lose certain cell structures and functions.

3.2.2 *Flagellates with hydrogenosomes*

The hydrogenosome is an organelle which occurs in a variety of anaerobic, but quite unrelated protists. Its main function is to ferment pyruvate, produced by glycolysis, into acetate and hydrogen; this process is coupled to energy conservation via substrate level phosphorylation (Müller 1980,

1988; see Section 3.3). They are defined and recognized by H_2-production (by whole cells or the isolated organelle) or by the cytochemical demonstration of the enzyme hydrogenase (Finlay and Fenchel 1989; Müller 1988; Zwart *et al.* 1988). Among the flagellates, hydrogenosomes are found in all trichomonads and probably always in the related oxymonads and hypermastigids. Among the trichomonads, one free-living form (*Pseudotrichomonas*) is known; otherwise they are commensals, occurring mainly in the intestinal tracts of animals. The two latter groups live as symbionts in the hindgut of wood-eating insects. These organisms have complex cells with an endoplasmic reticulum, dictyosomes and peculiar microtubular organelles.

The origin of hydrogenosomes has been debated (Cavalier-Smith 1987*b*, 1991*b*; Fenchel and Finlay, 1991*a*; Finlay and Fenchel 1989; Müller 1980, 1988, 1993). In the case of anaerobic ciliates and anaerobic chytrids, all evidence indicates that they are modified mitochondria (see below). In the case of the trichomonads, the hydrogenosomes are, as in ciliates, covered by a double membrane, but the inner membrane does not form cristae as is often the case in ciliates. The presence of DNA has not been demonstrated convincingly. Hydrogenosomes do not contain cytochromes, but they do have a pyruvate ferredoxin oxidoreductase and an hydrogenase. These are unusual enzymes in eukaryotes, but they are known in some anaerobic bacteria such as species of *Clostridium* (Section 2.1). Müller (1980, 1988) has therefore suggested that hydrogenosomes represent an analogy to mitochondria, but that they descended from a clostridial endosymbiont, rather than from an aerobic bacterium. On the other hand, the problem of explaining the presence of the particular enzymes also applies to the ciliates in which evidence for a mitochondrial nature of the hydrogenosomes is strong. Given the available evidence, we tend to agree with Cavalier-Smith (1987*b*, 1991*b*) that the hydrogenosomes of trichomonads (and their relatives) are also modified mitochondria. This is also supported by the finding that sequences of ferredoxin and of succinyl coenzyme A synthetase from *Trichomonas* hydrogenosomes are more similar to those of mitochondria than to those of anaerobic bacteria (Johnson *et al.* 1990; Lahti *et al.* 1992). A final answer will depend on a more detailed knowledge of the biochemical structure of the hydrogenosomal enzymes, or better still, on the sequencing of the organellar DNA (if it can be found).

The peculiar free-living, anaerobic flagellate *Psalteriomonas* (Broers *et al.* 1990, 1993) also has hydrogenosomes. It has been suggested that it belongs to the Heterolobosa, a group of otherwise mainly aerobic flagellates with mitochondria.

The taxonomically enigmatic anaerobe *Blastocystis*, which occurs in the human intestine, has morphologically typical mitochondria, but they do not contain cytochromes (Zierdt 1986). It is unknown whether they function as hydrogenosomes.

3.2.3 *Major protistan groups with some anaerobic representatives*

Several representatives of otherwise aerobic groups (heterolobosans, aplastidic chrysomonads, kinetoplastids and choanoflagellates) occur in anaerobic samples from sediments and can be grown in strictly anaerobic, crude cultures (Patterson and Fenchel, unpublished observations). These organisms have not yet been studied in any detail, but the situation resembles that of ciliates in that several forms seem to be facultative anaerobes or they have closely related, aerobic congeners.

The chytrids are common fungi-like organisms, most of which are aerobes. However, several anaerobic species have been isolated from the rumen of cows (Barr 1988; Munn *et al.* 1988) and it has been shown that their mitochondria-like organelles function as hydrogenosomes (Yarlett *et al.* 1986). The taxonomic position of the chytrids is not quite clear; they are traditionally classified among the oomycetes (water moulds) which are related to the chrysomonad flagellates, diatoms and some other groups within the so-called stramenopiles. According to Schlegel (1991) rRNA-sequencing of at least one species suggests that they are related to the real fungi.

In many respects ciliates are the best understood organisms with regard to the evolution of anaerobic forms. Ciliates are the most conspicuous eukaryotes in many anaerobic communities and some anaerobic forms have been studied in detail (see Sections 3.4–3.7). Due to their complex morphology, their systematics is relatively well understood. The traditional classification (Corliss 1979) and a revised version (based mainly on ultrastructural detail) by Small and Lynn (1981) is in many respects supported by the data emerging from rRNA-sequencing (Schlegel 1991; Fig. 3.18). However, with respect to the relation between the higher taxa, there are also some surprises and some of the classical orders seem to be polyphyletic. The order Heterotrichida provides an example: members of the genus *Metopus* are normally classified as typical heterotrich ciliates, but they seem to be quite unrelated to the other sequenced heterotrichs (Fig. 3.18).

The ciliates are basically aerobic organisms. The great majority of species are aerobes and an anaerobic lifestyle has evolved independently in many unrelated groups. This is already evident from Fig. 3.18. Corliss (1979) recognizes 22 orders of ciliates; as apparent from Table 3.2, anaerobic species occur in 11 orders, or in 8 orders if we exclude the karyorelictids, the hypotrichs and the prostomatids, which may include only facultative anaerobes. Among them, only the free-living odontostomatids and the rumen-dwelling entodiniomorphids are exclusively anaerobes. The odontostomatids, however, are probably derived from anaerobic heterotrichs and the entodiniomorphids from trichostomes (Corliss 1979; Jankowski 1964). In the other orders, aerobic species predominate. Among the heterotrichs,

Table 3.2. Ciliate orders with anaerobic representatives

Order	Hydrogenosomes	Remarks
Karyorelictida	–	Only some facultative anaerobes
Prostomatida	–	Few, with aerobic congeners
Haptorida	+	Few, with aerobic congeners
Trichostomatida	+	Several anaerobic families
Entodiniomorphida	+	All anaerobic
Suctorida	+	Few anaerobes
Scuticociliatida	+/–	Few, with aerobic congeners
Heterotrichida	+/–	Several anaerobic families
Odontostomatida	+/–	All anaerobic
Oligotrichida	– (?)	Only one species
Hypotrichida	–	Few species, perhaps only facultative anaerobes

the anaerobic habit has probably evolved at least once, since the Spirostomidae includes the anaerobic genus *Parablepharisma*. The Metopidae (which are apparently not related to the other heterotrichs) includes only anaerobes (genera: *Metopus*, *Brachonella*, etc.). At the moment the phylogenetic status of the Condylostomatidae is unclear; the family includes aerobic species, but also the anaerobe *Copemetopus*. The Caenomorphidae includes only anaerobes (*Caenomorpha*, *Cirranter* and *Ludio*), but this family is probably derived from the metopids. Finally, the nyctotherids and the clevelandellids include intestinal symbionts and are probably all anaerobes. The Trichostomatida is the other order which harbours many anaerobes. The plagiopylids (*Plagiopyla*, *Sonderia*, etc.) and the closely related trimyemids (*Trimyema*) are all anaerobes. The order also includes five families of anaerobic, intestinal commensals (the balantids which occur in the gut of a variety of invertebrate and vertebrate hosts including man, and the isotrichids, the paraisotrichids, the protoclavellids, and the blepharocorythids which all occur in the rumen or in the caecum of herbivorous mammals).

The remaining ciliate orders include fewer anaerobes. In most cases, individual species, which belong to genera with aerobic species, have adopted an anaerobic life (e.g. the genera *Cristigera* and *Cyclidium* among the scuticociliates and the genus *Lacrymaria* among the haptorids). Anaerobic suctorians (*Allantosoma* spp.) are known from the caecum of horses.

The order Karyorelictida seems to contain only facultative anaerobes, viz., members of the genus *Loxodes* which seems unique among the eukaryotes in that it performs NO_3^--respiration in the absence of O_2 (Finlay et al. 1983; Finlay 1985). The hypotrichids and the prostomatids have members (species of *Euplotes*, *Holosticha* and *Prorodon*) which can grow under anaerobic conditions (Fenchel and Finlay 1991*a*; unpublished observations), but these may not be obligate anaerobes.

All anaerobic ciliates have mitochondria or mitochondria-like organelles (Fenchel and Finlay 1991*a*), and aerobic ciliates which are tolerant of prolonged anoxia maintain their mitochondria with cristae although some morphological change is evident (Fenchel *et al.* 1989). In some cases the organelles of obligatory anaerobes are hydrogenosomes (Table 3.2; see also Section 3.4). In some taxonomic groups all members have hydrogenosomes (apparently all anaerobic trichostomes). Among the heterotrichs, the metopids and the caenomorphids have hydrogenosomes, but mitochondria in members of the genus *Parablepharisma* do not have hydrogenase. Most striking is the fact that anaerobic members of *Lacrymaria*, *Cyclidium* and *Cristigera* have hydrogenosomes, while their aerobic congeners have normal mitochondria. The scuticociliates have characteristic mitochondria which, during part of the life cycle fuse to form long organelles which stretch along the kineties from the anterior to the posterior end of the cell. The hydrogenosomes of *Cristigera* and *Cyclidium* have exactly the same unusual morphology as the mitochondria of aerobic congeners. It seems clear that mitochondria of some eukaryotic groups can easily change their normal function to that of pyruvate oxidation through H_2-excretion, and that this change has taken place independently, in many groups as an adaptation to anaerobic life. The mechanism by which the necessary enzymes were acquired is intriguing, but only a detailed biochemical study of the hydrogenosomes will resolve this question.

The role of non-H_2-evolving mitochondria in anaerobic protozoa is not yet clear, but it is certainly one of energy metabolism. Until further evidence is available, the most likely hypothesis is that they use some of the citric acid cycle enzymes for fumarate reduction to propionate and that this is coupled to electron transport phosphorylation. This is known from many invertebrates which can cope with anaerobic conditions (see Sections 2.2 and 3.5).

The ciliates of the rumen and caecum of herbivorous mammals are believed to have originated from free-living (anaerobic) ciliates which were introduced into the digestive tract (which, it must be assumed, already harboured anaerobic prokaryotes) through drinking water. An amazing variety of species evolved from these free-living ancestors. This must have happened in the period since the early to mid Tertiary and it represents an example of rapid adaptive radiation among microbes. A parallel is seen in the intestine of regular sea-urchins some of which harbour plagiopylid and metopid ciliates, which originally must have made their way into the gut with ingested marine sediments.

3.2.4 *How did anaerobic species evolve from aerobic ones?*

The immediate answer to this question is that the transition would seem to be only a small step. All living organisms, including the eukaryotes, which are essentially aerobes, carry an undeletable stamp of their anaerobic origin. This is particularly obvious in the surprisingly small role played by oxygen in essential biochemical processes, with the exception of oxidative phosphorylation (and in O_2-detoxification; see Section 3.3). To be sure, there are a few essential roles of oxygen in synthetic pathways (e.g. sterol synthesis; see Chapman and Schopf 1983), but since anaerobic habitats usually contain much eukaryote debris, the necessary compounds are probably available from the environment. By and large, any aerobic protozoon deprived of oxygen should be capable of fermentation. The energy yield per unit of food will be much lower, resulting in a slower growth rate, but since the organism will not be exposed to faster growing competitors it can colonize anaerobic habitats.

It is not our intention to discuss here all the constraints (which have ecological as well as physiological aspects) facing eukaryotes in anaerobic habitats or to discuss special adaptations for surviving anoxia; these are discussed in more detail in Sections 3.3 and 3.5. But the reason why most aerobic organisms are rapidly inactivated or killed if exposed to anaerobic conditions or to respiratory inhibitors such as CN⁻, is that the energy charge drops rapidly, thus blocking vital active ion-transport across the cell membrane.

Within some eukaryotic groups (including several invertebrates) there is wide variation with respect to survival to sudden anoxia or exposure to respiratory inhibitors. Organisms which are periodically exposed to anoxia in nature are capable of surviving this for shorter or longer periods. Many protozoa are microaerophilic: they depend on oxidative phosphorylation, but they are also sensitive to oxygen toxicity (Section 3.3) and they prefer, or only survive, at low oxygen tensions (Fenchel *et al.* 1989; Finlay *et al.* 1986). These organisms typically live in oxygen gradients and they are frequently exposed to anoxia (Section 5.1). These organisms are very tolerant of prolonged anoxia and exposure to respiratory inhibitors and they survive on the basis of fermentative metabolism.

Table 3.3 compares three ciliates: a strain of *Strombidium sulcatum*, which was isolated from plankton, which may be considered to be a fully aerobic species, and which grows well at high p_{O_2}; and the two microaerophilic species *Uronema marinum* and *Euplotes* sp. which prefer pO_2 values of ≈ 10 and $\approx 4\%$ atm. sat., respectively (the two latter species do, however, tolerate and grow at atm. sat.). It is seen that the *Strombidium* species is rapidly killed by anoxia and even by low concentrations of respiratory inhibitors. At the other extreme, the *Euplotes* species survives even high concentrations of CN⁻ and S²⁻ and is capable of infinite survival (or in some cases slow growth) under anoxic conditions.

Table 3.3.   Tolerance to anoxia and to respiratory inhibitors (at pH7.5) of three ciliates based on Fenchel *et al.* (1989) and unpublished results

Ciliate	Survival time[1]
Strombidium sulcatum	
Anoxia	5–10 min
S^{2-}	
5 mM	$\approx$ 1 min
0.2 mM	$\approx$ 40 min
CN$^-$	
2.5 mM	$\approx$ 20 min
0.5 mM	$\approx$ 80 min
0.1 mM	$\approx$ 3 h
Uronema marinum	
Anoxia	$\approx$ 6 h
S^{2-}	
5 mM	$\approx$ 2 h
0.5 mM	Infinitely[2]
Euplotes sp.	
Anoxia	Infinitely (slow growth in some cultures)
S^{2-}	
5 mM	$\approx$ 24 h
< 5 mM	Infinitely
CN$^-$	
2.5 mM	$\approx$ 2 h
0.5 mM	Infinitely (no growth)
0.1 mM	Growth

[1]Survival time means 50% survival (50% normal cell motility). [2]Means

In this section we have emphasized the evolution of obligate anaerobes which never had, or which have irreversibly lost, the capability of oxidative phosphorylation. The latter group probably evolved from microaerophilic ancestors. However, and as shown in Table 3.3, in some respects the distinction between aerobes and anaerobes is not sharp.

3.3   Energy metabolism and oxygen toxicity

So far we have discussed anaerobic energy metabolism only as it occurs in prokaryotic organisms. This metabolism is carried out in the cytosol, although in some cases it may be associated with the plasma membrane and

other internal membranes. In no case is the energy metabolism of prokaryotes localized within intracellular membrane-bounded organelles. Our
purpose here is to describe the principal features of energy-generating
metabolic pathways in anaerobic protozoa, to examine how intracellular
organelles modify this metabolism, to consider resemblances with aerobic
mitochondria, and to discover what, if anything, is uniquely different about
anaerobic energy metabolism in unicellular eukaryotes. It is unfortunate that
our discussion is hampered by incomplete information for the metabolic
pathways in free-living unicellular anaerobes. The best information available
has been obtained for parasitic and endocommensal species, especially
Giardia, the trichomonads, rumen ciliates and chytrids. However, we hope
to show that the metabolic pathways established within this group of
distantly related organisms reveal some fundamental similarities, and that
the additional limited evidence available does point to these similarities being
shared by the free-living species.

3.3.1 *Pathways*

Giardia lamblia and *Entamoeba histolytica* are both unicellular eukaryotes,
both parasites of man, and both capable of living in the complete absence
of oxygen. They have both been obtained in axenic culture; however, this
was achieved first for *G. lamblia* and therefore its metabolic pathways are
known in more detail. It is likely however, that the energy-yielding pathways
of *Giardia* and *Entamoeba* are broadly similar to each other. Neither
organism has mitochondria or hydrogenosomes, all of the enzymes of
metabolism are in the cytosol, there is no citric acid cycle, and no
cytochrome-mediated electron transport; and the basic energy-yielding
processes are fermentations.

Taking *G. lamblia* as the example — it is a flagellated protozoon which
infects a range of vertebrate hosts including man, attaching itself firmly to
the mucosal lining of the upper intestine where it typically experiences
conditions which are anoxic or nearly so. Glucose is the only sugar that is
metabolized, but the amino acid arginine is also an important energy source.
The basic fermentation pathway (pyruvate $\rightarrow$ acetate) is relatively simple
(Fig. 3.2) and similar to what we have already seen in fermenting prokaryotes
such as *Clostridium*. As in *Clostridium*, *G. lamblia* has pyruvate : ferredoxin
oxidoreductase (PFOR) activity, but note that this is not coupled to an
hydrogenase for the release of reducing equivalents as H_2-gas. Rather, it
simply dumps reducing power on organic compounds: in the case illustrated,
with the production of ethanol and alanine. There are no membrane-
bounded organelles in *Giardia*, although some recent work (Ellis *et al.* 1993)
has shown that the PFOR is associated with membranes, presumably the
plasma membrane or the endoplasmic reticulum. The iron-sulphur centre of

the ferredoxin in *Giardia* is of the [4Fe-4S] type, as it is in *Clostridium* (Müller 1993). Here we have a eukaryote which seems to have an energy-yielding metabolism and electron carriers which are not dissimilar to what we might find in anaerobic prokaryotes. But the parasite we have here is not an unaltered descendant of its ancestors from the pre-oxic world. One of the principal ways it has had to adapt (mainly because its hosts are aerobic organisms) is with respect to its periodic exposure to oxygen in the contemporary world. This is considered in more detail below. Recent reviews of research on *Giardia* can be found in Adam (1991) and Williams and Lloyd (1994).

Trichomonads too are parasitic flagellates although there are also a few free-living representatives, e.g. *Pseudotrichomonas* (Farmer 1993). *Tritrichomonas foetus* can cause abortion in cattle, and *Trichomonas vaginalis*, which is one of the best described species, is sexually transmitted between humans, living in the prostate and urethra of the male, and the vagina of the female. Carbohydrate is the main energy source of trichomonads, the metabolism is exclusively fermentative, there is no citric acid cycle and no cytochrome-mediated electron transport. Acetate, H_2 and CO_2 are the principal end-products, although lactate and ethanol can also be produced. Again, we find the pyruvate to acetate route at the core of the energy-yielding pathway, but this fermentation is fundamentally different from that in *Giardia*. Most obviously, the trichomonads release reducing equivalents as H_2 gas, and this activity takes place inside intracellular organelles known as hydrogenosomes. Furthermore, the ferredoxin which accepts electrons from PFOR is not the [4Fe-4S] version found in *Giardia* and *Entamoeba*, but a much larger molecule, with a [2Fe-2S] centre (Gorrell *et al.* 1984). There are also numerous other more subtle differences (see Müller 1992): for example, in trichomonads (as in mitochondria) the conversion of PEP to pyruvate is mediated by the enzyme pyruvate kinase, and by pyruvate: phosphate dikinase in *Giardia*.

Since their discovery by Lindmark and Müller (1973), hydrogenosomes have been found in a broad diversity of anaerobic eukaryotes (see also Section 3.2), including free-living anaerobic protozoa. Their main function is to ferment the pyruvate produced by glycolysis, into acetate and hydrogen; the process being coupled to energy conservation via substrate level phosphorylation. Here, we pay particular attention to the similarities of their function in different organisms.

Pyruvate produced by glycolysis is the typical substrate which enters the hydrogenosomes of trichomonads, rumen ciliates and anaerobic chytrids (Fig. 3.2). It undergoes oxidative decarboxylation, forming acetyl-CoA and thereafter acetate. The CoA moiety is transferred to succinate, forming succinyl-CoA which yields ATP by a substrate level phosphorylation. The reducing power from the oxidation of pyruvate is transferred to ferredoxin

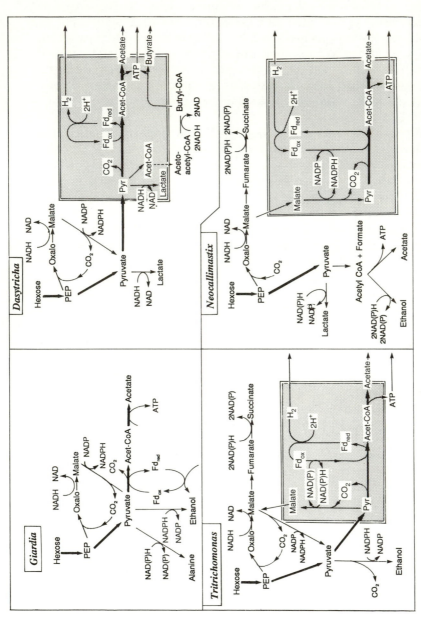

Fig. 3.2 Pathways of fermentation in four anaerobic eukaryotes. All except *Giardia* have intracellular redox organelles (hydrogeno-somes; shaded boxes). The hexose→pyruvate→acetate pathway (in bold), with some slight variations and attached ancillary pathways, plays a central role in the energy yielding metabolism of all four organisms. PEP, phosphoenolpyruvate; Fd, ferredoxin (based on many sources)

and then to protons, which, with the help of an hydrogenase, are reduced to H_2. Malate produced in the cytosol can also be a substrate if $NAD(P)^+$ is present, in which case it enters the hydrogenosome where it is oxidatively decarboxylated to pyruvate, the $NAD(P)H$ then being re-oxidized by reducing ferredoxin. Hrdý and Mertens (1993) could not find a pyruvate kinase in *Tritrichomonas foetus*, and they suggested that malate rather than pyruvate may be the principal substrate entering the hydrogenosomes in some trichomonads. These workers have made a special study of malic enzyme (decarboxylating malate dehydrogenase) which points to new roles for malate. They have proposed that in *T. foetus*, carbon flow is from PEP, to oxaloacetate to malate. The latter can then follow one of several routes (Fig. 3.2), e.g. conversion to fumarate followed by reduction to succinate (the same process found in prokaryotes, discussed in Section 2.2), oxidative decarboxylation to pyruvate in the cytoplasm, or as substrate for the hydrogenosomal malic enzyme. The last of these reactions is essentially reversible. Working in the pyruvate to malate direction (within hydrogenosomes), particularly under conditions of elevated CO_2 (Steinbüchel and Müller 1986a,b) the pathway may help preserve redox balance in the hydrogenosome and possibly also in the cytosol. For example, pyruvate : ferredoxin oxidoreductase and hydrogenase both have a redox potential of about -0.42 V which will allow evolution of H_2 even at relatively high pressures. But $NAD(P)H$ has an E_h of about -0.32 V, so $NAD(P)H$: ferredoxin oxidoreductase can donate electrons to ferredoxin only at very low H_2-pressures. However, if the pathway to H_2-evolution is inhibited by an elevated ambient H_2-pressure, ferredoxin can be re-oxidized by the same enzyme working in reverse, producing $NAD(P)H$, which is subsequently re-oxidized with the formation of malate, and the terminal production of succinate or ethanol. The evolution of such metabolic flexibility may indeed reflect the absence of an adequate H_2 sink in trichomonads. Support for this would be provided by the demonstration that similar multiple energy-yielding pathways were lacking from those anaerobic eukaryotes which do have efficient intracellular H_2-sinks such as intracellular methanogens (see Section 3.4).

Hrdý and Mertens (1993) have also found that the hydrogenosomal malic enzyme has a much higher affinity for NAD^+ than $NADP^+$ and that it probably uses NAD^+ preferentially. The same enzyme also operates in the mitochondria of anaerobic metazoa where it uses NAD^+ preferentially (references in Hrdý and Mertens 1993).

Dasytricha ruminantium is, as its name suggests, an endocommensal of ruminants. With respect to metabolic pathways, it is also the best described anaerobic ciliate. The energy-yielding pathways are broadly similar to those in trichomonads but there are two important differences. First, *D. ruminantium* can produce short chain fatty acids, especially butyrate from the condensation of two molecules of acetyl-CoA (Yarlett *et al.* 1985). The acetyl-CoA which is required for this is produced from pyruvate in the hydrogenosome although most of the steps towards butyrate formation (and the associated synthesis of ATP) take place in the hydrogenosome. The other main difference with trichomonads is that the malate dehydrogenase (decarboxylating, i.e. malic enzyme) is non-sedimentable and thus probably located only in the cytosol (Lloyd *et al.* 1989). There are also other differences; for example, lactate can be a final fermentation product in either the cytosol or the hydrogenosome (see also *Tetrahymena* below, in which lactate fermentation can occur in both the cytosol and inside mitochondria).

The anaerobic chytrid fungi play a significant role in the degradation of plant material in the rumen, where they secrete enzymes to hydrolyse plant structural polysaccharides. They are typical constituents of anaerobic fore-and hindgut microbial communities (Williams and Lloyd 1993), and it is likely that free-living species also exist. The best studied chytrid metabolism is that in the genus *Neocallimastix* which has a mixed acid fermentation (Yarlett *et al.* 1986; Marvin-Sikkema *et al.* 1993) with acetate, H_2, CO_2, formate, lactate and ethanol as end-products. Some succinate is also produced. The production of propionate or butyrate has not been reported. Detailed studies have been undertaken with *N. frontalis* and *N. patriciarum* (see references in Williams and Lloyd in press) and, most recently, with *Neocallimastix* sp. L2 (Marvin-Sikkema *et al.* 1993).

Neocallimastix produces H_2, CO_2, formate, acetate, lactate, succinate and ethanol during fermentation of glucose or cellulose. Unlike *Dasytricha*, malic enzyme is located solely in the hydrogenosomes rather than in the cytosol. Malate is converted to acetate and H_2 in the hydrogenosomes, and the pathway is similar to that in hydrogenosomes of other anaerobic eukaryotes. Acetate is also formed in the cytosol; conversion of pyruvate to formate plus acetyl-CoA is by pyruvate:formate lyase, and the acetyl-CoA so produced is converted to acetate (Yarlett 1994). So here we see the retention in a eukaryote of yet another prokaryote pathway (cf. Fig. 2.5). Note that relatively large quantities of acetyl-CoA may be produced in the cytosol. It is not known if this can be transferred from the cytosol into the hydrogenosomes. *Neocallimastix* sp. L2 lacks the citrate synthase and citrate lyase activity (Marvin-Sikkema *et al.* 1993) that would be required for this transport (see below).

Although no metabolic pathways have been described for free-living anaerobic eukaryotes it is likely that they are not very different from those in their commensal/parasitic relatives. The study by Goosen *et al.* (1990) is particularly relevant. They determined the end products of fermentation by the free-living hydrogenosome-containing anaerobic ciliate *Trimyema compressum* (the ciliate had, in the process of becoming established in laboratory culture, lost its endosymbiotic methanogens). The principal products were ethanol (which accounted for almost half of the excreted carbon), together with acetate, lactate, formate, CO_2 and H_2. Small amounts of succinate were detected, but no propionate or butyrate. Nothing is known about the metabolic pathways involved, but it is noteworthy that the end-products appear to be exactly the same as those produced by *Neocallimastix* (Fig. 3.2).

3.3.2 *Similarities between mitochondria and hydrogenosomes*

The basic job of the mitochondrion is to act as an electron sink (re-oxidizing NADH produced in glycolysis), and to produce ATP by electron transport phosphorylation. Hydrogenosomes and mitochondria both supply ATP to the cytosol. The hydrogenosome also acts as the primary sink for reducing power produced in glycolysis, and it too produces ATP, although probably only by substrate level phosphorylations. As we shall see in the next section (3.4), some hydrogenosome-containing cells do differ from those with mitochondria by also having a further electron sink — a symbiotic partner which is usually a methanogenic bacterium, and which consumes the H_2 produced by the hydrogenosomes.

It is worthwhile examining how mitochondria and hydrogenosomes actually acquire NADH in the first place, especially as the mechanisms involved are probably very similar. The NADH which is produced by glycolysis in the cytosol has to get inside the mitochondria or hydrogenosomes. Although the outer membrane of mitochondria is permeable to a wide range of solutes (most molecules with a molecular weight < 10 000), this is not true of the inner membrane, which is impermeable even to NADH. So, if the reducing power in NADH is ultimately transferred to O_2, how does it first get inside the mitochondrion? The solution is to use an 'electron shuttle', one of which is the malate shuttle (also known as the malate–aspartate shuttle: the biochemistry is quite complex and the full details need not concern us here). NADH in the cytosol, with the participation of malate dehydrogenase, reduces oxaloacetate to malate (also in the cytosol) which is able to cross the inner membrane (Fig. 3.3). In the matrix of the mitochondrion, the reducing equivalents which are carried by the malate are transferred to NAD^+, producing NADH with the help of matrix malate dehydrogenase. This reaction is shown for both mitochondria and hydrogenosomes in Fig. 3.3. In the mitochondrion, the reducing equivalents are eventually transferred to O_2, whereas in the hydrogenosome they reduce protons to H_2. The details of the malate shuttle may or may not be the same in both mitochondria and hydrogenosomes but it is undeniable that the transfer of malate does take place and that malate is the principal vehicle for the entry of reducing power into mitochondria, and into the hydrogenosomes of at least some organisms.

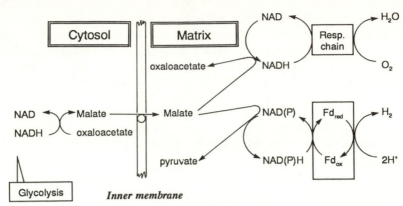

Fig. 3.3 The principal features of the malate shuttle as it is used to transport reducing equivalents (NADH produced in glycolysis) across the inner mitochondrial membrane. The reducing power passes into the respiratory chain and eventually reduces O_2 to H_2O. A similar shuttle may operate across the inner hydrogenosomal membrane, the reducing equivalents being transferred by ferredoxin (Fd) and eventually being released as H_2 gas.

In all aerobic organisms, Acetyl-CoA is used for fatty acid biosynthesis. This occurs only in the cytosol, but acetyl-CoA, which is formed from pyruvate, is produced only inside the mitochondria. So acetyl-CoA must somehow get out of the mitochondrial matrix and into the cytosol. The problem here is that acetyl-CoA, like NADH, needs some help to pass through the inner membrane. The way it does get out is known as the acetyl-group shuttle. It condenses with oxaloacetate to form citrate (which can pass through the inner membrane) in exchange for the import of malate into the mitochondrion: in other words the export of acetyl-CoA is linked to the import of reducing power and malate. The importance, or even the spread of occurrence of such a shuttle in anaerobic eukaryotes is unknown. In some parasitic organisms which do not synthesize their own fatty acids (e.g. the trichomonads; see Williams and Lloyd 1993), it may be irrelevant, and in *Neocallimastix* it may be unnecessary because acetyl-CoA is also produced by fermentation in the cytosol (Fig. 3.2). In *Dasytricha* however, we do apparently see some variation of the acetyl-group shuttle, as acetyl-CoA which is produced only inside the hydrogenosomes appears in the cytosol for the intermediate biochemical steps leading to the energy-yielding production of the volatile fatty acid butyrate in the hydrogenosome. Unfortunately the importance of malate as a substrate for *Dasytricha* is unknown.

The evolution of H_2-production in mitochondria is just one of the ways in which these organelles have adapted to anaerobic conditions. In Section 3.5 we will see how they sustain metazoa with an anaerobic energy-yielding metabolism; and in other unicellular eukaryotes which experience temporary anoxia we find a variety of adaptations and a remarkable metabolic flexibility within the mitochondria. Even in the ciliate *Tetrahymena*, which is typically aerobic, with a capacity for temporary life in the complete absence of oxygen, we see an apparent adaptation of the use of citric acid cycle enzymes within the mitochondria. It has long been known (Ryley 1952) that in the absence of oxygen *Tetrahymena* produces large amounts of succinate, smaller quantities of lactate and acetate, and it assimilates CO_2 (i.e. 2 glucose + CO_2 → 2 succinate + 1 lactate + 1 acetate). Incorporation of CO_2 is necessary for the formation of succinate; without it, lactate is the only fermentation product. Interestingly, some of the lactate dehydrogenase activity is located in the mitochondria of *Tetrahymena* (see Schrago and Elson 1980), as it also is in the hydrogenosomes of *Dasytricha*. *Tetrahymena* is basically an aerobe with a complete set of the citric acid cycle enzymes and the ability to reduce fumarate. Fumarate is used in *Tetrahymena* as an organic electron acceptor, and the reduction of fumarate is coupled to electron transport phosphorylation (see Rahat *et al.* 1964; Schrago and Elson 1980), probably in the same way as it is used in the mitochondria of 'anaerobic' metazoa (3.5).

There is also evidence that some unicellular eukaryotes may exploit inorganic terminal electron acceptors during extended periods spent in anoxic water. *Loxodes* is a microaerobic, free-living, freshwater ciliate. Its physiological ecology has been studied in detail and is reasonably well understood (Fenchel and Finlay 1984, 1986*a*,*b*; Finlay and Berninger 1984; Finlay and Fenchel 1986; Finlay *et al.* 1983, 1986; Finlay 1990). The ciliate has many unusual features: it is the only unicellular organism known to be capable of perceiving gravity; it is also sensitive to light, and it integrates its perception of blue light, oxygen, and gravity to quickly seek out its preferred habitat: darkness and a $p_{O_2} \leq 5\%$ atm. sat. The ciliate also spends long periods in anoxic water, indeed it is often the most abundant ciliate in the anoxic hypolimnion (Finlay 1981, 1985; Finlay *et al.* 1988) and it is likely that its success is partly related to its ability to switch to nitrate respiration. There is certainly a dissimilatory nitrate reductase in the cells and some circumstantial evidence that it is located in the mitochondria. The switch from aerobic to nitrate respiration is associated with a reduction of approximately 50% in molar growth yield because the electron transport chain to nitrate is shorter and has fewer sites for the phosphorylation of ATP (Fig. 2.19). So long as nitrate is not limiting, the energy deficit can be met by increasing the number of (shortened) chains — thus increasing the area of the inner mitochondrial membrane; and indeed, this is shown by an increase in surface area of about 70% (Finlay *et al.* 1983). *Loxodes* is probably not the only eukaryote with a capacity for switching to nitrate respiration. There is some evidence that other typically aerobic (and unrelated) ciliates living in anoxic lake water may also have dissimilatory nitrate reductases (Psenner and Schlott-Idl 1985; Hadas *et al.* 1992)

3.3.3 Coping with oxygen

Free oxygen presents special problems for anaerobes because some of the enzymes involved in energy metabolism (e.g. hydrogenases and PFOR) are inhibited by oxygen. As these organisms have no means of coupling oxygen uptake to energy conservation, continued energy generation by oxygen-sensitive fermentation pathways in the presence of O_2 will require either some ability to cope with the effects of oxygen, or some means of escaping exposure to oxygen.

Most anaerobes are periodically exposed to free oxygen, e.g. by living in the steep but fluctuating redox gradients of sediments, or in the variable oxygen tension of gastro-intestinal tracts. Their defences are both behavioural and biochemical. Behavioural adaptations are often easier to observe and interpret, especially in motile anaerobic eukaryotes such as ciliates. They include the transient (or 'phobic') response of rapid ciliary reversal as the cell swims from anoxia into a zone of detectable oxygen; or the typical kinetic response where swimming velocity increases with increasing oxygen tension, taking a minimum value in anoxia, thus 'trapping' cells in oxygen-free water (Fenchel and Finlay 1990*b*).

Anaerobes can also consume oxygen which enters their cells, by using it as a sink for reducing power. Most have this capacity for 'respiration', which may even approach the rate for similarly-sized aerobes (Fenchel and Finlay 1990*b*), but it is efficient only at relatively low oxygen tensions (the K_m is generally around the 1% atm. sat. value (Fenchel and Finlay 1990*b*; Lloyd *et al.* 1982); at O_2-tensions greater than about 2% atm. sat. the supply of reducing power is insuffient to cope with the flux of oxygen, the interior of the cell gradually becomes oxidized and the cells die. Oxygen consumption is easily seen in behavioural responses of populations (Fenchel and Finlay 1990*b*). When a population of anaerobic ciliates is placed in oxygenated water, the cells tend to clump together, each cell benefits from oxygen consumption by neighbouring cells, and all cells eventually benefit from the microscopic anoxic zone created by the population as a whole (Fig. 3.4). Oxygen consumption by the very large numbers of anaerobic ciliates in the rumen may also be important for the maintenance of a low oxygen tension there.

At the biochemical level, the business of transferring reducing power to oxygen is rather complex, both in the chemistry of the four electron reduction to $2H_2O$, and in the biochemical tricks by which this is safely achieved. Aerobic organisms manage extremely well because they have a beautifully adapted molecule, cytochrome oxidase, which initiates a cycle of steps by

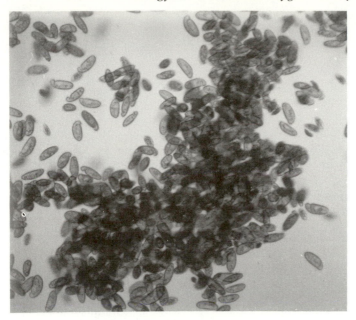

Fig. 3.4 Typical clumping of anaerobic ciliates when placed in oxygenated water. The cells consume oxygen and an anoxic microzone is eventually produced by the clump. The photograph shows a population of one ciliate species (*Plagiopyla frontata*) and each cell has a length of about 0.1 mm.

first fixing the O_2 molecule between Fe^{2+} and Cu^{2+} ions at the centre of the enzyme. None of the subsequent events in the carefully controlled stepwise reduction of oxygen release significant quantities of toxic intermediates.

However, anaerobes do not have cytochrome oxidase, and they face some problems in reducing O_2 to H_2O by a series of uncoordinated reactions. It is inevitable that toxic oxygen radicals and other compounds (e.g. singlet oxygen or hydrogen peroxide [H_2O_2]) will be formed along the way. Notable among the radicals is the superoxide anion, formed from the one-electron reduction of O_2:

$$O_2 + e^- \rightarrow O_2^-$$

and further reduced to hydrogen peroxide:

$$O_2^- + e^- + 2H^+ \rightarrow H_2O_2$$

Both products are potentially harmful and they are dealt with by two well-characterized enzymes; by superoxide dismutase (SOD) which catalyses the dismutation of O_2^-:

$$2O_2^- + 2H^+ \xrightarrow{\text{SOD}} O_2 + H_2O_2$$

and by catalase:

$$2H_2O_2 \rightarrow 2H_2O + O_2.$$
Catalase

In the real world however, the relevant biochemistry must be more complex. Flavoproteins, iron-sulphur proteins and a host of other electron transporters can all carry out the one-electron reduction of O_2, as can compounds such as humic substances in the surrounding water (Finlay *et al.* 1986). Sometimes the toxicity of oxygen is enhanced by known or unknown substances in the water or culture medium (Carlsson *et al.* 1978; Fenchel and Finlay, 1990*b*). There are also numerous well-characterized reactions between transition metal ions and partially reduced forms of oxygen, which produce very toxic intermediates, e.g. the production of the hydroxyl radical ($\cdot OH$) from the oxidation of ferrous iron:

$$Fe^{2+} + H_2O_2 \rightarrow Fe^{3+} + \cdot OH + OH^-.$$

Such a reaction could affect anaerobes living close to the oxic–anoxic boundary (where upwards diffusing ferrous iron is oxidized) in natural systems such as sediments and stratified water columns (Finlay *et al.* 1986; Davison and Finlay 1986). On the other hand, high concentrations of another transition metal ion (Mn^{2+}) appear to act as a substitute for the superoxide dismutase which is lacking in lactic acid bacteria (see Stanier *et al.* 1987).

There has been much debate about the nature and mechanisms of oxygen toxicity, and about the relevance of certain purported defence mechanisms (e.g. Baum 1984). Some things are undeniable: superoxide and hydrogen peroxide are certainly produced inside cells (Fenchel and Finlay 1990*b*; Finlay and Fenchel 1986; Finlay *et al.* 1986) but these may be only some of the toxic agents — merely fleeting intermediates (e.g. O_2^-) in the story. Enzymes such as SOD and catalase may or may not be important. It is true that neither SOD nor catalase are detectable in some strict anaerobes (e.g. *Clostridium* spp.) but the evidence from anaerobic unicellular eukaryotes hardly points to some universal strategy for dealing with oxygen toxicity. *Neocallimastix patriciarum* respires O_2 and produces both superoxide and peroxide. SOD is present but catalase is not detectable (Yarlett *et al.* 1987). In *Neocallimastix* sp. L2, neither SOD, catalase, nor peroxidases are detectable, and their role is probably substituted by NAD(P)H oxidase activity (Marvin-Sikkema *et al.* 1993), an enzyme better known in phagocytic white blood cells for catalysing the production of O_2^- used to kill ingested bacteria:

$$2O_2 + NADPH \rightarrow 2O_2^- + NADP^+ + H^+$$

The case of the anaerobic ciliate *Parablepharisma* is also enigmatic. It appears to be considerably more sensitive to oxygen than other free-living anaerobic ciliates (Fenchel and Finlay 1990*b*). It contains neither hydrogenase nor endosymbiotic methanogens and the target for inhibition by oxygen is unknown.

Anaerobic organisms obviously pay a price for never having had cytochrome oxidase, or for having lost it, but the details of the other methods they use to ameliorate oxygen toxicity are far from clear.

One thing which is clear however is that trace quantities of oxygen can be beneficial; probably providing a sink for reducing power so long as oxygen remains at a low level (about 1% atm. sat. or 3 μM — roughly the same as the apparent K_m for O_2-uptake (see Müller 1988). Oxygen consumption can balance inwards diffusion, so cells can retain a relatively constant, anaerobic interior. There is much evidence that this is generally the case with anaerobic unicellular eukaryotes which have hydrogenosomes, i.e. the free-living anaerobic ciliates *Metopus contortus* and *Plagiopyla frontata* (Fenchel and Finlay 1990*b*), rumen ciliates (see Williams and Lloyd 1993), trichomonads (Paget and Lloyd 1990) and anaerobic chytrids (Yarlett *et al.* 1987).

3.3.4 *Giardia*

Anaerobic protozoa which do not have oxygen-consuming hydrogenosomes must employ alternative biochemical strategies for dealing with periodic exposure to oxygen; and in *Giardia*, we find a remarkable ability to produce gradually more oxidized end-products in response to increasing p_{O_2}.

During the anaerobic fermentation of carbohydrates, *Giardia lamblia* forms four end-products: ethanol, CO_2, acetate and the amino acid alanine (Fig. 3.5). Anaerobically, ethanol is the principal product, whereas in the presence of oxygen, acetate and CO_2 become the major products. Alanine is produced only under anaerobic conditions. Even traces of O_2 (0.2% in N_2), producing a dissolved concentration below the limit of detection of 0.25 μM, partially oxidize the intracellular pool of NAD(P)H and stimulate the production of CO_2 and ethanol. A further increase in the oxygen pressure (to 0.8% in N_2; equivalent to about 3 μM dissolved oxygen) produces a more oxidized cytosol, further CO_2 production, a greatly enhanced rate of acetate production (ten times greater), but a return to the anaerobic level of ethanol production and a cut in alanine production rate to about one-sixth of the anaerobic level (Fig. 3.5). Thus, in response to a low but increasing oxygen tension, the carbon flux shifts from the production of alanine, to ethanol, to more oxidized end-products (acetate + CO_2). One might suppose that the mechanism underlying switching between fermentation pathways had something to do with the changing intracellular redox state, but it is not clear that this is so. The enzymes involved in the production of alanine from

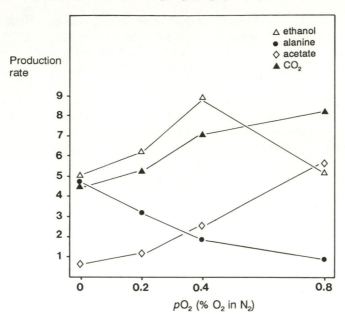

Fig. 3.5 End-product formation rates in the anaerobic flagellate *Giardia lamblia* at different oxygen tensions. Rates are nmol/min/10^7 trophozoites (based on data in Paget *et al.* 1993).

pyruvate with the concomitant oxidation of NAD(P)H (alanine amino-transferase and glutamate dehydrogenase) are not sensitive to oxygen; whereas the pyruvate : ferredoxin oxidoreductase (catalysing pyruvate → acetyl-CoA) *is* oxygen-sensitive; and the production of *both* ethanol and alanine involve the oxidation of NAD(P)H.

There is general acceptance that the trophozoites of *G. lamblia* living in the small intestine can experience variable oxygen concentrations, perhaps even in excess of 100 μM (see Williams and Lloyd 1993). It is also accepted that oxygen can be toxic (O_2-uptake by *Giardia* is maximal at 3.5–7 μM O_2 (about 2% atm.) and inhibited at 15–80 μM; Paget *et al.* 1989), and that the oxygen-tension controls the carbon flux through different metabolic pathways; but there is no agreement on the precise nature of the switching mechanisms.

Anaerobic eukaryotes also resemble prokaryotes in the mechanisms they use for dealing with the toxic effects of oxygen. All extant eukaryotes, both aerobic and anaerobic, are the products of a long period of evolution in a biosphere containing oxygen and all of them, including the anaerobes, seem to contain some means of protection against the toxic effects of oxygen. They all, for example, contain SOD, which disposes of the potentially dangerous charged superoxide radical (in contrast to the more variable occurrence of catalase and other enzymes which cope with peroxides which, having no charge, are more likely to diffuse out of the cell before reaching dangerous concentrations).

The extent to which aerobes and anaerobes produce protective enzymes varies, and it probably depends on the natural habitats concerned. Among the anaerobes, *Giardia* and *Entamoeba* have evolved a considerable tolerance of oxygen simply because they are frequently exposed to it by living in an aerobic host. For other anaerobes, such as *Parablepharisma*, exposure to anything greater than trace levels of oxygen is probably infrequent and they do not normally waste energy making protective enzymes for which there is no use. The obvious consequence is that such organisms appear particularly sensitive to elevated levels of oxygen provided in laboratory experiments. There is no evidence of any correlation between sensitivity to oxygen and the evolutionary age of eukaryotes: particularly sensitive forms are not relics from the pre-oxic world but forms which do not normally experience high levels of oxygen in their natural habitats.

Very few things appear to be uniquely different about the energy metabolism in unicellular eukaryotes: the broad spectrum of fermentation pathways found in prokaryotes is well-represented in eukaryotes and, with some notable exceptions (e.g sulphate reduction) respiratory electron transport processes also occur. Where we do find a difference is in the compartmentalization of energy metabolism within the hydrogenosomes, but again, the biochemistry of the H_2-evolving fermentations performed here is not very different to that performed in prokaryotes. With the assumption that hydrogenosomes have evolved from mitochondria, it is perhaps curious that their energy metabolism is based on fermentation rather than respiration. Two points should be made. First, the full metabolic repertoire of hydrogenosomes, especially in the free-living anaerobes, is unknown so it has never actually been disproved that these organelles may be capable of coupling ATP phosphorylation to electron transport processes such as fumarate respiration. Secondly, the widespread occurrence of H_2-evolving fermentations in anaerobic protozoa may be a consequence of the almost ubiquitous availability of hydrogenotrophs in anoxic habitats. The most obvious example is the widespread endosymbiosis of methanogens within anaerobic protozoa; and the reasons for believing that these symbioses may have favoured the evolution of H_2-evolving fermentations in anaerobic unicellular eukaryotes is discussed in Section 3.4.

3.4 Symbiosis with prokaryotes: intracellular syntrophy

One of the extraordinary things about the diversity of anaerobic protozoa is that most have other microorganisms living either inside them or attached to their external surfaces. These associations are usually symbiotic in so far as they are invariably permanent, they involve specific partners, and one or both partners 'benefit' from the assocation. It is true that there are exceptions: the diplomonad flagellates, for example, have neither mitochondria nor hydrogenosomes, nor (so far as is known) any other microorganisms living permanently associated with them. It appears as if their entire evolutionary history has passed without the emergence of functional associations with prokaryotes. But as a group, the anaerobic protozoa are remarkable for the high proportion of species which live in symbiotic association with prokaryotes. They are also unique in the types of symbiotic partners they have acquired: endosymbiotic purple non-sulphur photosynthetic bacteria and methanogens; and ectosymbiotic sulphate-reducing bacteria. One obvious feature which is common to all three of these prokaryote groups is that they can consume hydrogen as a substrate, and this points to the likely significance of H_2 production by the protozoa in the maintenance of these symbiotic associations. As we have already seen (Section 2.1), H_2-evolving fermentations are relatively productive in terms of the energy they yield but this productivity does depend on the maintenance of a low H_2-pressure. The problem, of course, gets worse as cells get bigger. Eukaryotes tend to be much bigger than prokaryotes but their greater size means that they can usually find room for symbionts. One could argue that because the efficiency of H_2-evolving fermentation (and hence the fitness of the eukaryote performing it) will be increased by having a resident sink for H_2, that evolution has favoured the retention of this type of functional consortium.

The point was stressed in Section 2.2. that the complete mineralization of organic matter by communities of anaerobic prokaryotes requires the integrated activities of several different forms (e.g. cellulose degraders, VFA producers, sulphate reducers and methanogens). Now, with the introduction of anaerobic eukaryotes we see a quantum leap in the simplicity and efficiency of the whole decomposition process. The protozoon comes with the capacity to engulf particles, and with a suite of digestive enzymes and a metabolic repertoire superior to anything found in the prokaryotes. The consequence is that the concerted activity of a variety of prokaryotes can be transferred to a eukaryote with a single prokaryotic symbiont species; and complex organic matter ingested by the eukaryote ends up as CO_2, plus either H_2S or CH_4.

3.4.1 *Phototrophic endosymbionts*

The first protozoon with H_2-consuming symbionts we will consider is also the most recently discovered (Fenchel and Bernard 1993*a,b*): the oligotrich ciliate *Strombidium purpureum*. Oligotrich ciliates are typical inhabitants of oxygenated waters; they are, for example, one of the dominant components of the microfauna living in the upper waters of the open ocean (Stoecker *et al.* 1987). Some species, however, are typically benthic, and this applies also to *S. purpureum*, described by Kahl (1930–35) and named according to the typically red to reddish brown colour of its cytoplasm. In other respects, *S. purpureum* is atypical. Its natural habitat is the illuminated zone of anoxic marine sandy sediment. It can be cultured in the complete absence of oxygen but only if it is periodically illuminated; and the purple colour in the cytoplasm is due to the several hundred photosynthetic non-sulphur bacteria which live in the cytoplasm of each ciliate (Fig. 3.6). These appear to be true endosymbionts, they contain photosynthetic membranes, and the study of their ultrastructure and photosynthetic pigments indicates that they are most closely related to *Rhodopseudomonas* (see Sections 2.2 and 3.1).

Some other clues from experimental work with cultured cells help us to understand how this consortium functions. The ciliate shows a photosensory behaviour, accumulating in anoxic water when illuminated, and migrating to the oxic–anoxic boundary (1–4% atm. O_2) in the dark. Furthermore, when illumination is provided as a spectrum of wavelengths, the ciliates accumulate at the wavelengths corresponding to the absorption peaks of the symbionts' photosynthetic pigments, indicating that the symbionts are physiologically active and that they have some influence over ciliate motility.

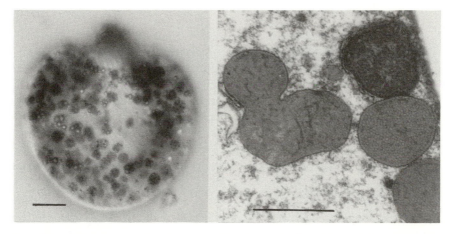

Fig. 3.6 *Left*: a living cell of the anaerobic oligotrich ciliate *Strombidium purpureum* showing the photosynthetic symbionts containing refractile food storage vacuoles. Scale bar represents 10 μm. *Right*: an electron micrograph of a thin section of *S. purpureum*, showing a mitochondrion (top right) and several symbionts (one of which is budding) containing photosynthetic vesicles and tubules (adapted from Fenchel and Bernard 1993*a*).

The current understanding of how the *S. purpureum* consortium functions, and its evolutionary significance, are as follows. All the close relatives of the ciliate, including its congeners, are aerobic organisms, so it is reasonable to suppose that *S. purpureum* became an anaerobic organism by evolution from an aerobic ancestor. The ciliate would then be similar to many others that have made the transition from aerobic to anaerobic environments, living by a H_2-evolving fermentation and doubtless also producing other electron sink products such as acetate and lactate (see Section 3.3) (although it should be pointed out that the evidence in support of H_2-evolution by the ciliate is still tentative; Fenchel and Bernard 1993*a,b*). It is proposed, however, that *S. purpureum* carried the process one step further, by acquiring photosynthetic symbionts capable of consuming the H_2, and possibly some of the other fermentation products, as reductants for photosynthesis. An additional benefit to the ciliate could be the transfer of some fraction of the symbionts' photosynthate: this is a well-known phenomenon in aerobic ciliates (see Finlay 1990) but its significance in *S. purpureum* is unknown. Neither is it known how the first ciliate host would have acquired its symbionts, although there are numerous examples of extant aerobic oligotrichs which ingest living photosynthetic eukaryotes, retrieving the chloroplast and retaining the latter as a functional intracellular photosynthesizing organelle (e.g. Stoecker *et al.* 1987). This capacity for sequestering photosynthetic activity does seem to be particularly well-developed in oligotrich ciliates.

However, the most interesting feature of the symbiosis concerns another characteristic of purple non-sulphur bacteria: in the dark and at low oxygen levels, they have a capacity for oxidative phosphorylation, using H_2 and fatty acids as substrates (see Section 2.3). This means that in the dark, the symbionts will benefit from access to oxygen and it explains the otherwise incomprehensible accumulation of anaerobic ciliates at low oxygen tensions in the dark (Fig. 3.7). Oxygen consumption by the symbionts presumably relieves the threat of oxygen toxicity on the ciliate and allows it, through this periodic conversion to microaerophily, to extend its range to the oxic–anoxic boundary. The *S. purpureum* consortium shows how the acquisition of an anaerobic symbiont with a capacity for aerobic respiration can be adaptive and can increase the fitness of an anaerobic host. It provides a clue to the type of symbiont which some ancestral amitochondriate eukaryote might have acquired, leaving unanswered only the question of how the host eventually sequestered the energy-yielding aerobic respiratory pathway for its own advantage, thus transforming the symbiont into an organelle.

Mitochondria are believed to be descended from endosymbiotic bacteria, and one of the most likely contenders for the bacterial ancestor is a form similar to a present-day photosynthetic non-sulphur bacterium (Woese

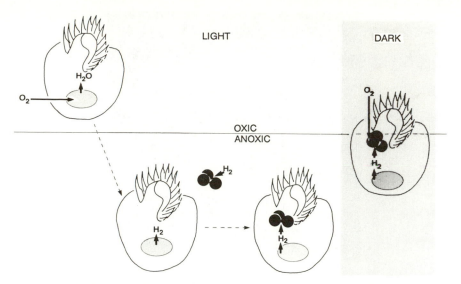

Fig. 3.7 Evolution and contemporary ecology of the *Strombidium purpureum* complex. The ancestral oligotrich ciliate (left) is believed to have been aerobic, with mitochondria which consumed oxygen. This ciliate evolved into an anaerobe with a H_2-evolving fermentation. Photosynthetic non-sulphur purple bacteria were acquired as endosymbionts which used the H_2 from the ciliate as reductant for photosynthesis; and through their capacity for respiration at low oxygen tensions in the dark, transformed the ciliate back into an aerobic organism.

1977). The *S. purpureum* consortium may provide a living analogy for the early stages in the evolution of mitochondria.

3.4.2 *Methanogen endosymbionts*

While taking a close look at the ciliates living in the anaerobic layers of marine sandy sediments, Fenchel *et al.* (1977) found that most had ectosymbionts, or endosymbionts, or both. It was some time before any of the ectosymbionts were identified (see ectosymbiotic sulphate reducers below) but the endosymbionts were generally easier to identify, especially those which had the characteristic autofluorescence of methanogens (Van Bruggen *et al.* 1983; Finlay and Fenchel 1989; Fenchel and Finlay 1991*a*).

Rumen ciliates were the first protozoa to be recognized with associated methanogens (Vogels *et al.* 1980; Krumholz *et al.* 1983) but the evidence was that they were attached only to the external surfaces of the ciliate. Very little is known about the mechanisms or reasons for this attachment. It is possible that the association is casual and the methanogens use the ciliates merely as a surface for growth. It is also possible that the methanogens use

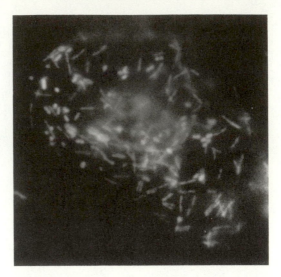

Fig. 3.8 Autofluorescing methanogenic bacteria (of at least two morphotypes) inside the ciliate *Dasytricha ruminantium* from the rumen of a sheep. Each methanogen is approximately 1 μm in length (adapted from Finlay *et al.* 1994).

substrate (e.g. H_2) diffusing out of the ciliates, but this has not been demonstrated and remarkably little is known about the identities and natural variation in abundance of these ecto'symbionts'.

It has recently been demonstrated (Finlay *et al.* 1994) that some of the ciliates living in the rumen of sheep also have intracellular methanogens (Fig. 3.8) which have all the hallmarks of endosymbionts (enclosed individually in membrane-bounded vacuoles, escaping digestion, and growing and dividing at more-or-less the same rate as the host ciliate). They are not present in all rumen ciliate species (e.g. the larger entodiniomorphids such as *Polyplastron*) but they are present in the typically very abundant small isotrichids (especially *Dasytricha ruminantium*) and the small entodiniomorphid *Entodinium* species (Fig. 4.25); in each of which they account for 1–2% of host ciliate biovolume.

Many other animals also have cellulose-rich diets. Notable among these are the many types of plant- and wood-eating insects which also have anaerobic microbial communities living in their hindgut (see 4.2). These break down the plant biomass into low molecular weight compounds which can be easily assimilated by the host. For reasons which are not entirely understood, the numbers of free-living methanogens and the pattern of electron flow in the hindguts of insects is biased towards H_2/CO_2-acetogenesis rather than H_2/CO_2-methanogenesis (see Section 2.1); the obvious benefit to the host animal being that it obtains acetate which can be oxidized as its

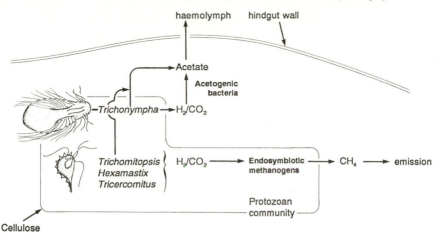

Fig. 3.9 An outline of the likely flow of carbon and reducing equivalents through the hindgut of a lower termite (e.g. *Zootermopsis angusticollis*). Cellulolytic flagellates produce H_2, CO_2 and acetate as fermentation products. Hypermastigid flagellates such as *Trichonympha* do not have endosymbiotic methanogens and the H_2 they produce is used by H_2/CO_2 acetogens; the acetate thus produced being transported across the gut wall for oxidation by the termite. Endosymbiotic methanogens in the small trichomonad flagellates (e.g. *Trichomitopsis*) consume H_2 produced by host metabolism (based on other similar schemes produced by Messer and Lee 1989 and Odelson and Breznak 1983).

principle source of energy (Fig. 3.9). The other interesting thing about these microbial communities is that they almost always include large numbers of protozoa, most of which probably have some cellulolytic activity; indeed it is likely that the protozoa living in the hindguts of lower termites are the principal agents of wood cellulose hydrolysis (Odelson and Breznak 1983; see also Section 4.2). Although the numbers of free-living methanogens in the insect hindgut are relatively low, termites and cockroaches do produce measurable amounts of methane. It is now known that this methane has its origin in the methanogens living inside the protozoa which are a part of these hindgut communities. For example, the methanogens living symbiotically inside the heterotrich ciliate *Nyctotherus ovalis* are the principal source of methane produced in the hindgut of the cockroach *Periplaneta americana* (Gijzen *et al.* 1991).

Of greater global significance perhaps, is the methane produced by termites, which also has its origin in endosymbionts. Yamin (1978) obtained a culture of the cellulolytic flagellate *Trichomitopsis termopsidis* from the hindgut of the 'lower' termite *Zootermopsis*. The culture was claimed to be axenic, but Odelson and Breznak (1985) discovered that it produced methane and that this activity could be eliminated with the specific methanogen inhibitor bromoethanesulfonic acid. Lee *et al.* (1987) then proceeded to show that the same termite host (*Z. angusticollis*) contained in its hindgut several species of trichomonad flagellates (*Trichomitopsis termopsidis, Tricercomitus termopsidis* and *Hexamastix termopsidis*) all of which demonstrated the characteristic autofluorescence of methanogens and, in electron micrographs, apparently endosymbiotic bacteria (which, we now know, resemble verified endosymbiotic methanogens from other protozoan species). It is likely that all termite flagellates have cellulolytic activity, but not all of these flagellates have endosymbiotic methanogens. They seem to be restricted to the trichomonads and they are absent, for example, from the much larger hypermastigids, such as *Trichonympha* (Fig. 3.9). Only the lower termites (the families Masto-, Kalo-, Hodo- and Rhinotermitidae; including the species *Reticulitermes flavipes* and *Zootermopsis angusticollis*) contain hindgut microbial communities in which protozoa are an integral part. The 'higher' termites contain only bacteria in their hindgut.

Confirmation of the existence of endosymbiotic methanogens in free-living protozoa was obtained relatively recently (Van Bruggen *et al.* 1983) and was followed by a steady increase in the diversity of known protozoan hosts. Although the search for methanogens in free-living ciliates has been pursued with some vigour, this cannot be said for other protozoa, so our knowledge of the true diversity of hosts is incomplete. Our impression that methanogens are absent from free-living anaerobic flagellates may simply reflect the limited diversity examined so far (although there are unpublished observations, some of them our own, of methanogen autofluorecence in free-living trichomonads). Endosymbiotic methanogens are however definitely present in some protozoa which lack hydrogenosomes, e.g. the rhizomastigids *Pelomyxa palustris* (Van Bruggen *et al.* 1988) and *Mastigella* spp. (Van Bruggen *et al.* 1985). *Pelomyxa palustris* is particularly interesting. It appears to contain three morphologically distinct types of rod-shaped endosymbionts; two of which show characteristic methanogen autofluorescence, while the third, the 'thick-type' endosymbionts, have an axial cleft, no autofluorescence, and they surround the nuclei of the host. With the assumption that at least one of the methanogen types in *Pelomyxa* consumes H_2, and the recognition that *Pelomyxa* has neither hydrogenosomes nor any other obvious H_2-producing organelles, it has been suggested (Van Bruggen *et al.* 1988) that the thick-type symbiont might perform the H_2-producing role of the hydrogenosomes. If this is true (and *Pelomyxa* is particularly unyielding of its secrets) then the interior of a *Pelomyxa* would actually represent a three-step food chain (e.g. eukaryote fermentation to VFAs → prokaryote H_2 production → methanogenesis), but there is no experimental evidence for any of this. *Mastigella* also contains 'thick-type' endosymbionts (which resemble those in *Pelomyxa*) as well as endosymbiotic methanogens (Van Bruggen *et al.* 1985).

Table 3.4. Symbiotic bacteria associated with anaerobic free-living ciliates from marine and freshwater habitats

Species	Habitat	Endosymb. rhodobacteria	Endosymb. methanogens	Ectosymb. bacteria
Cristigera (3 spp.)[1]	Marine			+
Paranophrys sp.	Marine			
Metopus (3 spp.)[2]	Marine		+	+[12]
Metopus 'major'	Marine		+	?
Metopus verrucosus	Marine		?	+
Metopus sp.	Marine		+	
Caenomorpha levanderi	Marine		+	+[12]
Caenomorpha capucina	Marine		?	+
Parablepharisma (4 spp.)[3]	Marine			+
Spathidium sp.	Marine			
Strombidium purpureum	Marine	+		
Holosticha fasciola	Marine			
Plagiopyla frontata	Marine		+	−/+
Plagiopyla ovata	Marine		+	
Sonderia (8 spp.)[4]	Marine			+
Trimyema (2 spp.)[5]	Marine			
Saprodinium halophila	Marine		+	
Myelostoma bipartitum	Marine			−/+
Prorodon sp.	Marine			
Lacrymaria (2 spp.)[6]	Freshwater/Marine		+	
Brachonella spiralis	Freshwater		+	
Metopus (4 spp.)[7]	Freshwater		+	
Caenomorpha (3 spp.)[8]	Freshwater		+	
Bothrostoma undulans	Freshwater		+	
Tropidoatractus acuminatus	Freshwater		+	
Parapodophrya denticulata	Freshwater		+	
Saprodinium (2 spp.)[9]	Freshwater		+	
Plagiopyla nasuta	Freshwater		+	
Trimyema (2 spp.)[10]	Freshwater		+	
Holophrya sp.	Freshwater		+	
Cyclidium porcatum	Freshwater		+	
Isocyclidium globosum	Freshwater		+	
Isocyclidium globosum	Freshwater		+	+[13]
Caenomorpha (2 spp.)[11]	Freshwater		+	+[13]
Saprodinium difficile	Freshwater		+	+[13]
Epalxella oligotricha (*E. striata*)	Freshwater		+	
Holophrya bicoronata	Freshwater		+	

The ciliates listed belong to at least eight Orders, i.e. Scuticociliatida (e.g. *Cyclidium*), Heterotrichida (e.g. *Metopus*), Haptorida (e.g. *Spathidium*), Stichotrichida (e.g. *Holosticha*), Plagiopylida (e.g. *Plagiopyla*), Odontostomatida (e.g. *Saprodinium*), Prostomatida (e.g. *Holophrya*), Suctorida (*Parapodophrya*)

[1]*C. vestita, cirrifera,* sp. [2]*M. contortus, vestitus, halophila.* [3]*P. collare, pellitum, chlamydophorum,* sp. [4]*S. vorax, tubigula, sinuata, schizostoma, mira, vestita, cyclostoma, macrochilus.* [5]*T.* sp. a, b. [6]*L. elegans, sapropelica.* [7]*M. palaeformis, es, striatus, setosus.* [8]*C. uniserialis, corlissi, sapropelica.* [9]*S. mimeticum, dentatum.* [10]*T. compressum,* sp. [11]*C. medusula, lata.* [12]The ectosymbionts of *M. contortus* and *C. levanderi* are known to be sulphate reducers (Fenchel and Ramsing 1992). [13]From a sulphate-rich solution lake (Finlay *et al.* 1991; Esteban *et al.* 1993*b*)

The best studied anaerobic amoeboflagellate is the heterolobosan described and named *Psalteriomonas lanterna* by Broers *et al.* (1990). It does contain hydrogenosome-like organelles and they form tight complexes with the methanogens. This is a phenomenon which is particularly common in the free-living anaerobic ciliates (Fenchel and Finlay, 1991*c*; Esteban *et al.* 1993*a*; Finlay *et al.* 1993), the group for which our knowledge of methanogen endosymbiosis is most extensive, and the group to which we now turn our attention.

The diversity of known free-living anaerobic ciliates belong to at least eight ciliate orders and there are representatives with endosymbiotic methanogens in all of them (Table 3.4). With respect to these higher level taxa there is no difference between the marine and freshwater fauna (apart from the presence only in the latter of the anaerobic suctorian *Parapodophrya*); and even at the genus level there are very few differences (*Sonderia* and *Parablepharisma* have not been found in freshwater; see Section 4.2). It also appears as if a greater number of anaerobic ciliate species have evolved in freshwater (see Fenchel and Finlay 1991*a*; Section 4.2). But the most obvious difference is illustrated in Table 3.4; endosymbiotic methanogens have been recorded from all the freshwater ciliates on the list, but from less than one-third of the marine species. Moreover, more than half the marine species have ectosymbionts, some of which have recently been demonstrated to be sulphate reducers (see below). The only anaerobic freshwater ciliates with ectosymbionts are those recovered from water with an unusually high sulphate concentration.

Before we attempt to explain these differences it is necessary to try and understand why methanogens live inside protozoa. As we shall see later, not all endosymbiotic methanogens are closely related to each other and it is possible that their functional significance is not identical in all cases. Nevertheless, they all probably fulfil one common basic function by acting as an additional electron sink (Fig. 3.10), thus compensating for the limited oxidative capacity of the hydrogenosomes (at least in comparison with mitochondria). The following section explores the role of endosymbiotic methanogens as electron sinks.

3.4.3 *Methanogens and intracellular* H_2

Methanogens live inside anaerobic protozoa because there they find a virtually guaranteed supply of H_2 produced by the hydrogenosomes (see Fig. 3.11). The benefits enjoyed by the protozoa are not quite so obvious, but they are measurable in terms of their influence upon host growth rate and yield. These benefits are unlikely to have anything to do with 'milking' or digesting the symbionts: it is readily calculated that complete utilization

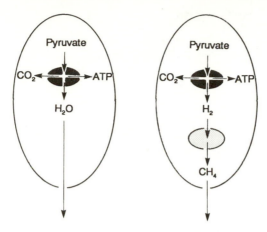

Fig. 3.10 Functional similarities between organisms with mitochondria and hydrogenosomes. The organism on the left is aerobic, with mitochondria. Pyruvate enters the mitochondria where it is decarboxylated, ATP is produced and reducing equivalents are released as water. The process is essentially similar in the anaerobic, hydrogenosome-bearing eukaryote on the right; but the ATP yield is much lower and in many cases, an additional electron sink is present — in the case shown, an endosymbiotic methanogen which releases CH_4.

of symbiont production would account for an increase in host growth rate of < 1% (Finlay and Fenchel 1993).

In the two large anaerobic ciliates *Plagiopyla frontata* and *Metopus contortus*, experimental inactivation of the methanogen endosymbionts decreases ciliate growth rate by 25–30% (Fenchel and Finlay 1991*b*) (Fig. 3.12). In the much smaller anaerobic ciliate *Metopus palaeformis* (Finlay and Fenchel 1991*a*), endosymbiotic methanogens have only little impact upon the growth of the host ciliate. These limited data tend to suggest that the larger ciliates in particular may grow better when they harbour methanogens.

A possible role for the intracellular H_2-tension has been demonstrated both experimentally and theoretically. An experiment was performed with the large anaerobic ciliate *Metopus contortus*, using bromoethanesulfonic acid (BES), a compound which is a specific inhibitor of methanogens. The rate of H_2-production by cells treated with the inhibitor was plotted against the rate of CH_4-production by untreated cells (Fig. 3.13). As 4 mols of H_2 are required to produce 1 mol of CH_4, a line with a slope of 1/4 was superimposed on the figure. If, following inhibition of the methanogens, all H_2 that was previously consumed by the methanogens was now released from the ciliates, the data points should lie on the line. Clearly, this is not the case. There are several possible explanations for the apparent reduction in H_2-evolution. The ciliates could simply switch to formate production for example. In the absence of any further experimental data, one explanation

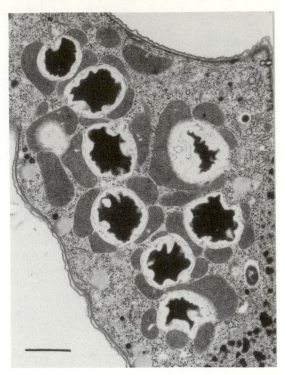

Fig. 3.11 Intracellular methanogens (the electron dense particles) surrounded by hydrogenosomes in the anaerobic ciliate *Trimyema* sp. Scale bar = 1 μm.

which does at least have some theoretical support is that inactivation of methanogens leads to an increased p_{H_2} within the ciliate, followed by partial inhibition of H_2-production. Considering a spherical ciliate with dimensions close to those of *Plagiopyla* (radius 25 μm), which produces 12 pmol H_2/ciliate/hour (Fig. 3.14), the calculated p_{H_2} at the surface of the ciliate will be about 10^{-2} atm. (Fenchel and Finlay 1992). This is roughly 1000 times the environmental p_{H_2}, and much higher than 10^{-4} atm.: the pressure at which many H_2-yielding oxidations become exergonic. It is thus possible that hydrogen transfer is a significant component of the symbiosis between ciliates and methanogens.

3.4.4 *Endosymbiosis as a refuge*

There are some other reasons for the methanogens living inside protozoa. Free-living methanogens fare badly in competition with sulphate reducers, and they are rapidly inactivated by exposure to oxygen. Endosymbiosis within protozoa alleviates both problems; there are probably no endosym-

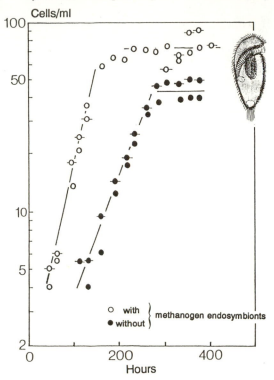

Fig. 3.12 Population growth of the anaerobic ciliate *Metopus contortus* in culture. Growth rate and yield are both significantly lower in ciliates without endosymbiotic methanogens (adapted from Fenchel and Finlay 1991*b*).

biotic sulphate reducers, and oxygen consumption by the host maintains intracellular anoxia. The latter is confirmed by the observed retention of methanogen autofluorescence, and about half the original methane generation rate when the ciliates *P. frontata* and *M. contortus* are exposed to low ambient oxygen tensions (< 2–5% atm. sat.; Fenchel and Finlay 1990*b*, 1992).

3.4.5 *The ciliate as a natural 'chemostat'*

It is likely that endosymbiotic methanogens are completely dependent on substrate supplied by the host, and that their growth rate is dictated by the host growth rate. This must be relevant to the observation that the volume fraction of methanogens remains relatively constant at 1–2%. In an attempt to explain this approximately constant volume fraction we have recently (Finlay and Fenchel 1992) drawn on the analogy of the chemostat as a model

mol **CH₄** ciliate⁻¹h⁻¹ x 10⁻¹²

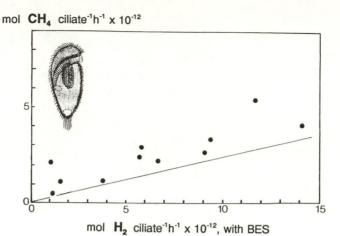

mol **H₂** ciliate⁻¹h⁻¹ x 10⁻¹², with BES

Fig. 3.13 Methane production by cells of *Metopus contortus* containing endosymbiotic methanogens, plotted against H₂-production by cells in which the methanogens have been inhibited using BES. The data points all lie above the line which shows the expected relation if all H₂-production normally goes to the methanogens and the rate of H₂-production is unaffected by inhibiting the methanogens (adapted from Fenchel and Finlay 1992).

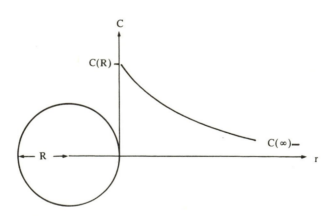

Fig. 3.14 Diffusion of H_2 from a hypothetical spherical ciliate. The ciliate has radius R, and H_2 is produced with the rate P and removed from the cell by diffusion. $C(\infty)$ is the bulk H_2-concentration and $C(r)$ the concentration at distance r from the centre of the cell. It can be shown (Fenchel and Finlay 1992) that $C(R)$, the concentration at the surface of the cell without H_2-consuming symbionts equals $P/(4\pi RD) + C(\infty)$. For a ciliate with $R = 25\,\mu m$, producing H_2 at a rate $P = 12$ pmol/h, taking $D = 10^{-5}$ cm²/s, and the solubility of H_2 as 1.6 mg/l, the H_2-pressure at the surface of the ciliate is 0.013 atm. (= 1.3 kPa).

for the apparent steady state. To a degree, the population of intracellular methanogens behaves as it would in a chemostat vessel, with the substrate (H_2) being provided by host metabolism, and the 'dilution rate' being the rate of increase in ciliate cell volume. Thus, applying a basic equation of chemostat theory:

$$dx/dt = x\mu - Dx$$

where μ, x, t and D are, respectively, substrate-dependent growth rates of endosymbionts, endosymbiont population density, time and dilution rate (= rate of growth in host cell volume): μ will decrease with increasing x until a steady state is reached (i.e. $dx/dt = 0$, and $\mu = D$).

However, this model does require some modification because a growing ciliate is not quite identical to a chemostat in the way it functions. Although the methanogens are diluted at a rate which is exactly equivalent to the host growth rate, some of their substrate (H_2) will diffuse out of the ciliate before they can capture it. It is inevitable that some substrate will be lost, at a rate which is proportional to ciliate cell surface area and to the internal substrate concentration. The mechanism of dilution in a true chemostat ensures that the microbial population and the unused substrate are both diluted together, and at the same rate: it does not allow some selective leakage of unused substrate, with simultaneous retention of the entire cell population. However, general chemostat theory can be adapted to account for the apparent stability of endosymbiont population densities; this is developed in Box 3.1.

We can apply the adapted chemostat model to the growth of endosymbiotic methanogens in the ciliate *Plagiopyla frontata*, incorporating data given in Fenchel and Finlay (1992), i.e. ciliate maximum growth rate constant $(G) = 0.03$ h^{-1}; Y of methanogens (assuming 2.8 g dry wt/mol CH_4, and dry wt as 20% of wet wt) = 3.5 ml/mol H_2; α (maximum H_2-production rate)/($G_{max} \times$ cell volume), i.e. $(2 \times 10^{-11}$ mol H_2/ciliate/h)/ $(0.03 \times 70\,000$ μm^3) = 0.0095 mol H_2/ml; F (flux/cell volume), i.e. $(4\pi RD_c/80\,000$ μm^3) = 16159 h^{-1} (assuming $R = 25$ μm; $D_c = 0.036$ cm^{2}h^{-1}); $K = 5 \times 10^{-9}$ mol H_2/ml; μ_m (max. growth rate of methanogens) = 0.11 h^{-1}. Substitution into eq. [8] in Box 3.1 gives a predicted equilibrium symbiont biovolume $(\bar{x})$ of 0.03 (i.e. it accounts for 3% of host ciliate biovolume, which is reasonably close to the figure of 2% obtained independently by measurement of symbiont and host ciliate volumes (Fenchel and Finlay 1992). So long as $G << \mu_m$, the last term in eq. [8], remains very small and the equilibrium values for methanogen abundance biomass $(\bar{x})$ are sensitive only to the methanogen yield coefficient (Y) and the specific substrate production rate (α). Similarly, in [7], where $G << \mu_m$, the equilibrium H_2 pressure $(\bar{s})$ is sensitive only to growth rates and K. Thus the number of methanogens and the H_2-pressure change little at low ciliate growth rates. Only in the rather unrealistic circumstances of G approaching μ_m, would $\bar{x}$ and $\bar{s}$ both change

Box 3.1   The host–endosymbiont system as a chemostat

Assuming Monod kinetics, the expression for the rate of change in bacterial population density in a chemostat is:

$$dx/dt = x\mu_m [S/(S + K)] - Dx \qquad [1]$$

where μ_m is maximum bacterial growth rate; S, substrate concentration; D, dilution rate; and K is the substrate concentration at which $\mu = \frac{1}{2}\mu_m$. The expression can be adapted very simply to apply to the growth of endosymbiotic methanogens in a ciliate:

$$dx/dt = x\mu_m [S/(S + K)] - Gx \qquad [2]$$

where $G (= dV/dt \cdot 1/V)$, is the volume (V) - specific growth rate of the host ciliate, i.e. $\mu_{ciliate}$.
But the description of the changing substrate concentration in a real chemostat is:

$$dS/dt = DS_r - DS - \mu x/Y \qquad [3]$$

where S_r is reservoir substrate concentration and Y is cell yield, and it must, in the case of the endosymbiotic methanogens, be adapted to account for the loss of substrate across the cell surface of the ciliate:

$$dS/dt = G\alpha - FS - \mu x/Y \qquad [4]$$

where α is the quantity of S produced per unit growth of host cell, and F is a constant, accounting for diffusion of substrate out of the host cell ($= 4\pi R D_c$/cell volume; where R is host cell radius and D_c is the diffusion coefficient of the substrate).
    At steady state, $dx/dt = dS/dt = 0$, and the steady state substrate concentration $(\bar{s})$ and population density $(\bar{x})$ are:

For bacteria in a chemostat:		For endosymbiotic methanogens:	
$\bar{s} = DK/(\mu_m - D)$	[5]	$\bar{s} = GK/(\mu_m - G)$	[7]
$\bar{x} = Y(S_r - DK/(\mu_m - D))$	[6]	$\bar{x} = Y(\alpha - FK/(\mu_m - G))$	[8]

dramatically (Fig. 3.15), as they would also do in a chemostat operating at high dilution rates (i.e. $\bar{x}$ would decrease rapidly to zero leading to 'wash-out', and $\bar{s}$ would increase quickly towards the 'reservoir concentration'). The interesting conclusion here is that in this type of symbiosis, where the endosymbionts are completely dependent on a host metabolite, the association is automatically stabilized.

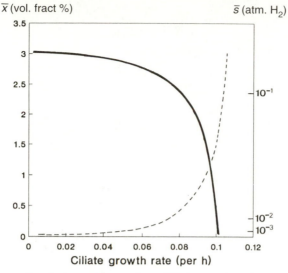

Fig. 3.15 The predicted dependence of equilibrium values for endosymbiotic methanogen volume fraction ($\bar{x}$; solid line) and hydrogen pressure ($\bar{s}$) on the growth rate of the host ciliate. See text for values substituted into eq. [7] and [8]. Note that for all realistic ciliate growth rates (≤ 0.03 h^{-1}), $\bar{x} \leq 3\%$ and $\bar{s} \leq 10^{-3}$ atm. H$_2$/CO$_2$ methanogens have a potential to grow at rates of about 0.35 h^{-1} (original).

3.4.6 *The identity and evolution of endosymbiotic methanogens*

In the last few years the electron microscope (EM) has revealed a wealth of information about the ultrastructure of the methanogens inside ciliates (see Finlay and Fenchel 1993; Embley and Finlay 1994). This information has provided the first insight into processes such as polymorphic transformation which have later been verified by modern molecular methods of identification. The EM studies also revealed a variety of levels in the apparent sophistication of the physical relationship between methanogen and host organelles. In *Metopus palaeformis*, for example, the methanogens are always rod-shaped (Fig. 3.16) and they do not seem to be attached to the hydrogenosomes. But the symbiosis in *Metopus contortus* is more complex. Rod-shaped methanogens do live in this ciliate, but they coexist with a variety of other forms which are stages in the progressive stripping of the cell wall: a process which culminates in the attachment of the transformed symbionts to hydrogenosomes (Finlay and Fenchel 1991*b*). Finally, the methanogen symbiosis in *Plagiopyla frontata* might be viewed as the most advanced evolutionary stage described so far. The symbionts are all disc-like forms, attached to hydrogenosomes, usually in stacks of alternating hydro-

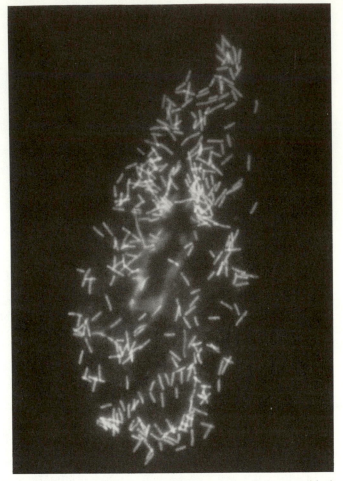

Fig. 3.16 Autofluorescing rod-shaped methanogens living symbiotically inside the anaerobic ciliate *Metopus palaeformis*. Most methanogens are 2–3 μm in length (original).

genosomes and methanogens, and each stack behaving as a single, integrated organelle (Fig. 3.17). The activity and life cycle of the symbionts in this ciliate also seems to be under the control of the ciliate. When the ciliate divides, the hydrogenosomes and symbionts in the stacks also divide, and in a synchronized fashion, which ensures the integrity of future stacks and a constant number of stacks in succeeding generations. Is there any evidence that this is indeed an evolutionary progression of forms: a case of evolution bringing increasing sophistication in the degree of interaction with the eukaryote host? This would mean that there may be some co-evolution between host and symbionts, and then more questions arise; e.g. can we

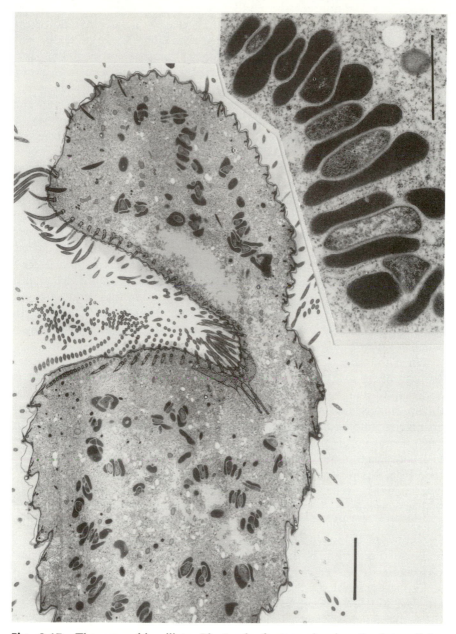

Fig. 3.17 The anaerobic ciliate *Plagiopyla frontata* showing the intracellular complexes of alternating hydrogenosomes (electron dense) and symbiotic methanogens. Scale bars are 10 μm (whole ciliate) and 1 μm (higher magnification inset, top right).

establish the identities of the endosymbionts and are they identical to free-living forms? Do different methanogen species, or merely morphological variants of the same species, coexist inside each ciliate?

Following the traditional microbiological approach of isolation and cultivation, a number of Dutch workers obtained cultures of rod-shaped methanogens from washed anaerobic protozoa. Using various characteristics of the phenotype of this cultured material they have, in the cases of the ciliates *Metopus striatus* (Van Bruggen *et al.* 1984) and *Plagiopyla nasuta* (Goosen *et al.* 1988), the giant anaerobic rhizomastigid *Pelomyxa palustris* (Van Bruggen *et al.* 1988) and the flagellate *Psalteriomonas vulgaris* (Broers *et al.* 1993), identified the symbiont as the methanogen *Methanobacterium formicicum*. The conclusion to be drawn from these studies is that a single methanogen species has colonized and established endosymbiotic associations with a diverse collection of unrelated anaerobic protozoa. As the methanogens concerned appear to be identical or very similar to each other and to the free-living form of *M. formicicum*, their establishment as endosymbionts would appear to have occurred in the recent past, or else their long symbiotic existence has occurred without significant divergence from the free-living phenotype. The single exception in these culture-based studies is the marine ciliate *Metopus contortus*, from which Van Bruggen *et al.* (1986) isolated an apparently new methanogen, which they named *Methanoplanus endosymbiosus*.

When the same questions concerning symbiont identity are addressed using the methods of molecular biology, a radically different perspective is obtained. Certain base sequence motifs in rRNAs are unique to each of the three domains eubacteria, archaebacteria and eukaryotes, and it is now possible to amplify an rRNA from one domain without interference from organisms from the other domains in the same sample. Thus, it is possible to amplify endosymbiotic methanogen rRNA without competitive interference from the host ciliate. The base sequence in the rRNA can then be determined, a specific oligonucleotide probe can be constructed, tagged with a fluorochrome, and introduced to the original protozoan–methanogen consortium. If the specific probe hybridizes with (and highlights) the symbiotic methanogen and that alone, there is little doubt that the highlighted organism is identical to that from which the original sequence was obtained.

Embley *et al.* (1992*a*,*b*), Embley and Finlay (1994), and Finlay *et al.* (1993) used these procedures with five different ciliate–methanogen consortia. Only a single archaebacterial sequence was obtained from each consortium. As each consortium contained only *intra*cellular particles with characteristic methanogen autofluorecence, and as the consortia produced methane (Fenchel and Finlay 1992), the inescapable conclusion was that each of the five ciliate species contained a single methanogen species. This corroborated the postulated polymorphic transformation of a single methanogen species in *Metopus contortus* (Finlay and Fenchel 1991*b*); which was later conclusively demonstrated using a specific probe to the *M. contortus* symbiont (Embley *et al.* 1992*a*).

The first sequence was obtained from the symbiont in *Metopus palae-formis* (Embley *et al.* 1992*b*). It shares some similarities with sequences from other free-living methanogens in the Methanobacteriales lineage (Fig. 3.18) but it is obviously not identical to any known free-living methanogen and it could, if there was good reason, be given its own species name. The same could also be said for the methanogen in *Metopus contortus* — it is similar to, but still distinctly different from the free-living *Methanocorpusculum parvum*. The marine ciliate *Metopus contortus* and the freshwater ciliate *Trimyema* sp. are not closely related to each other (Fig. 3.18), yet their symbiotic methanogens are quite closely related. Interestingly, the symbionts in these last two are also phenotypically similar to each other, for they both undergo polymorphic transformation inside their respective host ciliates.

In the case of the two *Metopus* consortia investigated, the hosts are closely related to each other but the symbionts are not. The same also applies in the case of two *Plagiopyla* consortia; the symbiont in *P. frontata* has *Methanolobus* as its closest free-living relative whereas the symbiont in *P. nasuta* forms a deep branching lineage within the Methanomicrobiales. Thus, each sequence recovered from these anaerobic cilates is 'novel' in so far as it is similar to but distinctly different from any known free-living methanogen. Closely related ciliates can have unrelated methanogens, distantly related ciliates can have methanogens which are quite similar to each other; so these symbiotic associations have formed independently and repeatedly.

With one exception (the rRNA-sequence obtained for *Methanoplanus endosymbiosus* confirms it to be closely related to the free-living *Methanoplanus limicola*) these studies of genotypic and phenotypic diversity of methanogen endosymbionts give different overall impressions. The rod-shaped methanogen in *Metopus palaeformis* is in the same major methanogen lineage as *Methanobacterium formicicum*, but the rod-shaped symbiont in *Plagiopyla nasuta* is not. Sequences obtained from an archived sample of symbionts from *Metopus striatus* and *Pelomyxa palustris* are indeed closely related to *M. formicicum* but it is curious that the sequences from the archive material are absolutely identical to each other.

Figure 3.18 shows phylogenetic trees based on rRNA-sequence data for the anaerobic ciliates and their methanogen symbionts. There is no congruence between the two phylogenies, and there is clear evidence that in *Metopus* and *Plagiopyla* at least, symbioses must have become established after the speciation of the hosts. As symbionts have obviously been acquired independently from different lineages of methanogens, there is no support for the idea that the endosymbiosis in *Plagiopyla frontata*, for example, is a

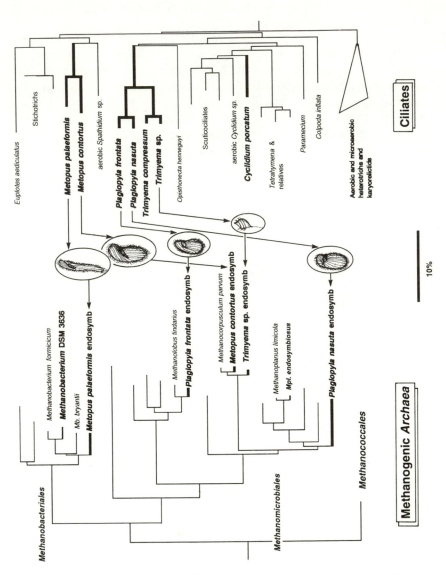

Fig. 3.18 Phylogenetic trees for ciliates and methanogens based on small subunit rRNA sequences. The branches shown in bold represent the anaerobic ciliates and endosymbiotic methanogens which have recently been sequenced. DSM 3636 = Deutsche Sammlung von Mikroorganisms strains 3636 and 3637 (adapted from Embley and Finlay 1994). The scale bar represents 10% estimated base substitutions.

highly evolved version of the symbiosis in *Metopus palaeformis*. Rather, the sophisticated association in *P. frontata* must be the product of co-evolution between this specific host and its symbiont.

3.4.7 Ectosymbiotic sulphate reducers

Most research on anaerobic protozoan consortia in the last few years has concentrated on the role of symbiotic methanogens but it is clear that these are just one of the prokaryote groups involved. A variety of anaerobic ciliates, in particular those living in marine environments, are known to support ectosymbiotic bacteria (Fenchel *et al.* 1977; Fenchel and Finlay 1991*a*). These attach to the ciliates by various means. The bacteria on the outer surface of *Parablepharisma pellitum* are inserted in pits in the cell membrane of the ciliate (Fig. 3.19), whereas in *Metopus contortus* and *Caenomorpha*, the bacteria live in sheaths which attach to the host cell membrane (Fig. 3.19). In *Sonderia*, they lie embedded in a thin mucous layer which covers the ciliate surface. Only recently (Fenchel and Ramsing 1992) has the process of revealing the identity of these bacteria begun, and again, modern molecular techniques have proved themselves invaluable. When the marine ciliates *Metopus contortus* and *Caenomorpha levanderi* were treated

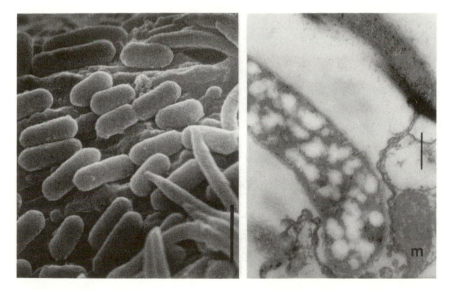

Fig. 3.19 Bacteria attached to the external surfaces of two anaerobic ciliates. *Left*: *Caenomorpha* sp. from a sulphate-rich lake (see Finlay *et al.* 1991), scale *bar = 1* μm. *Right*: the marine ciliate *Parablepharisma pellitum*. The bacterium is anchored in a pit in the ciliate cell membrane and a mitochondrion (m) is close by. Scale bar = 0.1 μm.

Isocyclidium globosum

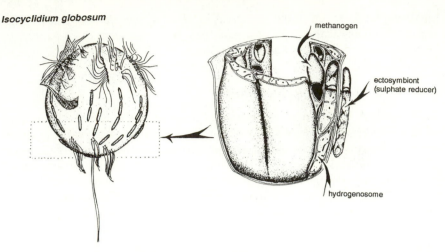

Fig. 3.20 *Isocyclidium globosum*, a small (20 μm) anaerobic ciliate from sulphate-rich freshwater. It has hydrogenosomes with cristae, endosymbiotic methanogens, and ectobiotic bacteria which may be sulphate reducers, although this has not been verified (adapted from Esteban and Finlay in press).

with an oligonucleotide probe for sulphate reducers (complementary to some parts of the 16S rRNA molecule which are specific to known free-living sulphate reducers belonging to the δ-group of the purple bacteria), the bacteria attached to the ciliates gave a positive response. There is also circumstantial evidence that they are sulphate reducers and that they use substrates (e.g. H_2, acetate, lactate, ethanol) diffusing out of the ciliate. First, they are abundant only on ciliates living in sulphate-rich waters, e.g. seawater, and sulphate-rich freshwater lakes (Finlay *et al.* 1991). Secondly, they may compete with intracellular methanogens for common substrates. In marine ciliates (e.g. *Metopus contortus*) with endosymbiotic methanogens the latter account for 1–2% of host biovolume, and the ectosymbionts for an estimated 0.4% (Fenchel and Ramsing 1992). This is probably true also for freshwater ciliates in sulphate-rich environments (Fig. 3.20). The ectosymbionts become significantly more abundant only on cells which do not have endosymbiotic methanogens (e.g. *Sonderia vorax* and *Parablepharisma*). On *Parablepharisma*, the ectosymbionts may account for 10–15% of host biovolume. If this is produced (as seems likely) at the expense of substrates diffusing out of the ciliate, then the bacteria concerned must have an anaerobic metabolism with a higher energy yield than anything known for endosymbiotic methanogens (Fenchel and Finlay 1992); thus it is likely that they are sulphate reducers. The fact that the ectosymbionts on *Parablepharisma* did not respond positively to the sulphate reducer probe is not conclusive proof that they are not sulphate reducers. It is possible that they are sulphate reducers, with a ribosomal RNA which is sufficiently different to prevent hybridization with the probe that was used.

It is not known if sulphate reducers can be endosymbionts. A guaranteed supply of inorganic electron acceptor might be a problem, but this could be circumvented by adopting fermentative pathways on a long-term basis. It is just possible that other endosymbionts which are known to be eubacteria are sulphate reducers, e.g. the bacteria living in the stable association with hydrogenosomes and methanogens inside the ciliate *Cyclidium porcatum* (Esteban *et al.* 1993*a*).

3.4.8 *Endosymbiotic methanogens — global significance*

Anaerobic protozoa are found in all but the most transient of anoxic habitats. They can be relatively abundant; free-living anaerobic ciliates typically reach several tens per ml, or $> 10^3$ per cm^2 of sediment (Fenchel *et al.* 1990; Finlay 1982; Finlay *et al.* 1991; Guhl *et al.* 1994), and rumen ciliates often number $> 10^5$ per ml (Hobson 1989). The question of the relative importance of endosymbiont methanogenesis on a global scale is frequently posed, and experimental results obtained in the last few years help greatly when trying to answer the question, at least for the free-living species which we will consider first. The relevant data are (a) the rate of methane production by individual symbiotic methanogens (which at 20 °C seems to be relatively constant at 1×10^{-15} mol CH$_4$ per hour for growing ciliates (Fenchel and Finlay 1992), and (b) the number of methanogens which actually live inside protozoa. The appropriate arithmetic is completed by multiplying by the abundance of the protozoa.

The few data available for the abundance of anaerobic ciliates in freshwater sediments lie within the range 500–2000 cm^{-2} (Finlay 1982; Wagener *et al.* 1990; Guhl *et al.* 1994). Taking account of the species in the ciliate community and the methanogen complement per ciliate; and having measured methane production by the complete sediment cores, we have calculated that symbiotic methanogens are responsible for a maximum of 5% of total sediment methane production. Roughly the same answer is obtained if we construct a schematic representation of a methanogenic system with endosymbiont-bearing phagotrophs (Fig. 3.21; Fenchel 1993). All anaerobic yields are assumed to be 10% and it is assumed that all fermenting bacteria are eventually consumed by the phagotrophs. Some of the assumptions in the model are perhaps a little unrealistic but the figure of 3% obtained for the endosymbiont contribution to methanogenesis does indicate that the theoretical and the experimental approaches are more or less concordant: endosymbiont methanogenesis is relatively unimportant in freshwater sediments and in methanogenic systems in general (i.e. where sulphate is absent or scarce).

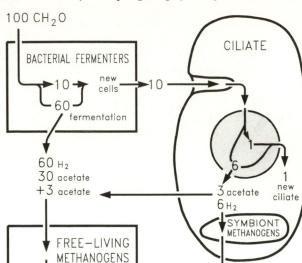

Fig. 3.21 Schematic representation of carbon flow in a methanogenic system. The anaerobic community includes anaerobic protozoa which feed principally on the fermenting bacteria. The protozoa contain symbiotic methanogens, but because the latter depend on the protozoa, which themselves represent an additional step in the food chain, methanogenesis from endosymbionts is a relatively small amount, being approximately 3% of the total. Cell yield (gross growth efficiency) is assumed to be 10%, and 30% of absorbed or ingested material is assumed to be lost or egested. Methanogenesis is calculated according to $CH_3COOH \rightarrow CH_4 + CO_2$, and $4H_2 + CO_2 \rightarrow CH_4 + 2H_2O$.

The picture is quite different in anoxic marine sediments, where fermentation products are finally mineralized by sulphate reducers (which outcompete free-living methanogens; see Section 4.1) rather than methanogens. In such habitats, anaerobic ciliates with symbiotic methanogens can be relatively abundant, especially in localized patches of decomposing detritus, and although total methane production may be relatively small, endosymbionts may account for most of it. In a decomposing heap of *Zostera* in a marine bay, Fenchel (1993) recorded 5100 anaerobic ciliates cm^{-2}; total methanogenesis was 4.1 mmol m^{-2} day^{-1}, and of this, the methanogen symbionts in ciliates contributed 87%.

The anoxic layers in marine sediments can eventually become depleted of sulphate. This allows the free-living methanogens to increase in abundance, and endosymbiont methanogenesis, as a proportion of the total, decreases to a figure typical of any methanogenic system.

The rumen on the other hand is a special type of methanogenic system, for the phagotrophs do not necessarily represent an additional step in the food chain. Many of the protozoa participate directly in digesting and fermenting the macerated plant parts which enter the rumen, and the free-living fermenters in the rumen simply do not have access to much of this material. This direct grazing activity probably contributes to the relatively high ciliate biomass in the rumen; it also means that rumen ciliate endosymbionts, acting at the end of a relatively short food chain, could be relatively more important than their relatives living inside free-living ciliates. We have recently shown (Finlay *et al.* 1994) that each small ciliate living in the sheep rumen typically contains several hundred methanogens. Assuming typical rates of methanogenesis per endosymbiont, this multiplies up to the equivalent of about 37% of the 19 litres CH_4 (Moss 1992) produced each day by a typical sheep. It remains to be seen if endosymbiotic methanogens are found in ciliates living in other ruminants. If so, it is just possible that the phenomenon has some global significance. Methane is the most abundant hydrocarbon in the troposphere and the emission from ruminants, at 75–100 Tg yr^{-1} (Crutzen 1991), is probably the second largest anthropogenic source, after rice cultivation (Hogan *et al.* 1991). The loss of methane from ruminants also has a more direct economic significance for it represents part of the animals' total energy intake (which partly explains the widespread enthusiasm for 'defaunating' ruminants; Leng 1991).

Termites also have not escaped the notice of global warming watchers and various investigators have estimated that termite emissions may account for anything between 5 and 40% of global CH_4-production (see refs in Brauman *et al.* 1992). Unfortunately, it is even more difficult to count termites than ruminants; and this, coupled with a plethora of termite feeding types (wood, grass, soil, fungus-growing) and methanogenic activities, does make it rather difficult to make robust estimates. The global significance of termite flagellates and their endosymbionts is not easily dismissed, but neither have they passed the audition for the global warming bandwagon.

3.5 Anaerobic metazoa

Anaerobic metazoa are probably non-existent in terms of the definition of anerobes given in Section 1.1 (i.e. organisms which can complete their life cycle in the absolute absence of oxygen). The existence of such anaerobic animals has been claimed repeatedly, but the evidence is largely circumstantial or anecdotal, and a metazoan without the capability of oxidative phosphorylation is yet to be found. On the other hand, many small sediment-dwelling metazoa, as well as intestinal parasites, do seem to be capable of growth and normal behaviour for extended periods in the absence of oxygen or to be capable of completing their life cycle at an extremely low p_{O_2}. These animals may play a role in anaerobic communities so it is appropriate to include a discussion of metazoa in the context of this book. Metazoan life without oxygen has recently been reviewed in Bryant (1991); a classical review is that of Brand (1946).

It is customary to distinguish between 'functional' and 'environmental' anaerobiosis. A typical example of the former is the anaerobic metabolism of muscles during exercise (running, burst swimming in fish, etc.) when the transport of oxygen cannot cope with the demand. Functional anaerobiosis is typically localized in some particular tissue. The metabolic pathway is glycolysis, with lactate or 'opines' (see below) as metabolites. This allows for a high power output and the metabolites are excreted, or metabolized when the tissue is re-supplied with oxygen. We will not consider functional anaerobiosis further here.

Environmental anaerobiosis occurs when the animal finds itself in anaerobic conditions. Animals which are normally aerobic, may at regular or irregular intervals be exposed to anoxia for short periods (e.g. lugworms at low tide, inter-tidal barnacles or mussels which close their shells during exposure to air to prevent dessication). Animals may also be trapped in anoxic water for shorter or longer periods (e.g. small benthic invertebrates on the bottom of stratified lakes or fish in ice-covered ponds). Such organisms are adapted to survive transient periods of anoxia. In contrast, some meiofauna species and intestinal parasites may depend on a steady state anaerobic energy metabolism for most of the time. There is a continuum between these extremes, and different types of metabolic pathways and other adaptations vary accordingly. Species which are adapted to survive long periods of anoxia and those which are largely dependent on anaerobic metabolism usually possess higher energy-yielding fermentative energy metabolism as a supplement to glycolysis. This takes place in the mitochondria and involves some of the enzymes of the citric acid (TCA-) cycle together with electron transport phosphorylation. Organisms which are regularly exposed to anoxia tend to store glycogen and to metabolize it during anoxia since this yields one additional mole of ATP per glucose during glycolysis.

Neither anaerobic respiration involving inorganic electron acceptors, nor pyruvate oxidation based on H_2-excretion, are known from metazoa. The main fermentative pathways which have been established are shown schematically in Fig. 3.22. The glycolytic pathway may lead to pyruvate which is then reduced to lactate in order to restore the cytoplasmic redox balance. Alternatively, pyruvate may be de-carboxylated prior to reduction which leads to ethanol and CO_2 as metabolic end-products. In marine annelids and molluscs in particular, an amino acid is co-metabolized during pyruvate reduction leading to 'opines' (strombine, octopine) as metabolic end-products rather than lactate. The opines are neutral and they are osmotically

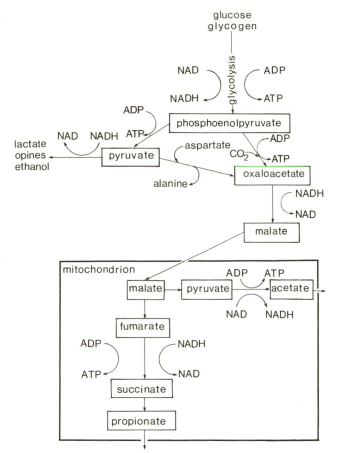

Fig. 3.22 The most important pathways of anaerobic energy metabolism in metazoa. For further explanation, see text (compiled from various sources).

less active than lactate; they can therefore be stored, to be metabolized once oxygen is available again (deZwaan and Putzer 1985; Schöttler and Bennet 1991). Pyruvate may also be co-metabolized with aspartate leading to alanine (which is not metabolized further) and oxaloacetate. The oxaloacetate is then reduced to malate, which is metabolized in the mitochondria (see below).

Most crustaceans, which are periodically exposed to anoxia, seem to rely only on the glycolytic pathway (Zebe 1982, 1991). Crustaceans are generally sensitive to hypoxia and many species do not tolerate anoxia even for short periods. Many other kinds of invertebrates, which are exposed to anoxia, also rely on the glycolytic pathway during the initial phase of anaerobiosis. Thus when the p_{O_2} falls, the rate of glycolysis increases accordingly so that the energy charge (ATP/[AMP+ADP+ATP]) is maintained. Glycolysis allows for a high rate of power production, but it is also wasteful in terms of substrate consumption (cf. discussion in Section 2.1). During prolonged anoxia, animals depend on the mitochondrial metabolism of malate, a process which increases the ATP-yield per mole glucose, but it is also associated with a lower rate of power production. In many normally aerobic invertebrates and vertebrates, the long-time strategy for anaerobic survival is associated with a decreased power output which may involve complex adaptations and coupling with membrane transport mechanisms (Hochachka 1986). The level of activity with respect to feeding and motility also decreases and the animals become lethargic. Anaerobic mitochondrial metabolism has been found in different kinds of annelids and molluscs. It is also found among intestinal commensals and parasites (nematodes, cestodes, trematodes and acanthocephalans) with a steady state anaerobic metabolism (Barrett 1991; deZwaan 1991; deZwaan and Putzer 1985; Schöttler 1977; Schöttler and Bennet 1991; Schroff and Zebe 1980). It must be assumed that such metabolic pathways are also operational in some meiofauna species and in some anaerobic protozoa (see Section 3.3).

Some metabolic features are probably common to all forms of anaerobic mitochondrial energy metabolism. Phosphoenolpyruvate deriving from glycolysis is carboxylated by CO_2-fixation to form oxaloacetate. This is reduced to malate, thus restoring the cytoplasmic NAD/NADH ratio. Malate enters the mitochondria and undergoes a branched fermentation leading to succinate (or propionate) and acetate as metabolites. The exact pathways vary somewhat, but they maintain mitochondrial redox balance and they result in additional ATP-generation. Part of the malate is transformed to fumarate which is reduced to succinate. This process is coupled to membrane electron transport phosphorylation and involves an iron-sulphur protein, a quinone and cytochrome *b* (see also Section 2.2). Succinate may be further de-carboxylated to propionate which is an alternative end-product. A fraction of the malate pool is oxidized to acetate via pyruvate and the resulting reduction equivalents are used for fumarate reduction. The energy of the pyruvate oxidation is conserved through substrate level phosphorylation (cf. 2.2 and Fig. 2.14). These pathways use some of the normal mitochondrial enzymes, and a reversal of part of the citric acid cycle. However, the properties of the fumarate reductase have in some cases been shown to differ from succinate dehydrogenase of aerobic tissue with respect to the affinities for succinate and fumarate (Barrett 1991).

We will not further discuss the biology and metabolism of the variety of larger, free-living aquatic animals which are occasionally exposed to and capable of surviving anoxia; this is treated in detail in Bryant (1991). We will, however, discuss two groups of animals in some more detail: intestinal parasites and the meiofauna of sediments. In both cases these animals may be a component of anoxic communities and they may be capable of living most of their life sustained by anaerobic energy metabolism. With respect to the meiofauna species which are found in anaerobic and reducing sediments, the degree of dependency or independency of oxygen remains an open question. Finally, we will discuss the somewhat related topic of sulphide toxicity and sulphide detoxification.

3.5.1 *Inhabitants of the intestinal tract*

The intestinal tract of animals, and herbivores in particular (which typically contain substantial amounts of undigested material which is degraded by microbes) constitute an anaerobic or almost anaerobic habitat. Those tracts therefore harbour obligate anaerobes including fermenters and methanogens and often anaerobic protozoa as well (Sections 2.2 and 4.2). Nevertheless, the intestine is probably rarely completely anoxic since some oxygen will diffuse through its lining, creating microaerobic conditions close to the wall (Czerkawski and Cheng 1989; Hillman *et al.* 1985). Representatives of some lower invertebrate groups (cestodes, digenean trematodes, nematodes, acanthocephalans, together referred to as 'helminths') have also adapted to life in the guts of animals. Although these organisms are not entirely independent of oxygen, they do represent the best studied examples of metazoan adaptation to anoxic life; they are also in many ways unusual due to their parasitic mode of life.

The energy metabolism of the parasitic helminths has been studied and reviewed by Barrett (1981, 1984, 1991). Some species depend almost exclusively on glycolysis, with lactate as the principle metabolite, but most forms have an '*Ascaris*-type' metabolism which involves carboxylation of phosphoenolpyruvate through CO_2-fixation and fumarate reduction as described above (Fig. 3.22). The metabolites of these forms are therefore mainly lactate, succinate, propionate and acetate in various proportions, but some species also excrete other compounds such as malate, butyrate, isobutyrate and valerate. The worms depend on carbohydrates (glycogen, glucose) as substrate and there is no co-fermentation with amino acids.

All studied species have a complete citric acid cycle (although some of the enzymes have a low activity) and a complete electron transport system for oxidative phosphorylation. When exposed to air, the worms respire, but the rate of fermentation continues almost unchanged, usually with the same products and the fraction of organic carbon degraded through the citric acid cycle is usually < 10%. The respiration is cyanide-insensitive which suggests that there is an alternative oxidase in addition to cytochrome *c* oxidase and it is possible that part of the oxygen uptake is not coupled to energy conservation, but functions as a detoxification mechanism (cf. Section 3.3). In some cases, exposure to atmospheric oxygen adversely affects motility, survival and growth, but the parasitic helminths are all capable of thriving for extended periods under strictly anaerobic conditions. However, all seem to require at least periodic exposure to some oxygen in order to complete their life cycle.

The role of oxygen in synthetic reactions is often cited as a barrier to complete independence of oxygen in eukaryotes. In addition to steroid synthesis and fatty acid desaturation and some other reactions, metazoa also use oxygen in quinone tanning and especially in the synthesis of collagen. The parasitic helminths are incapable of most oxygen-dependent synthetic reactions including steroid synthesis and quinone tanning in the adult. But the latter process does occur in the eggs released by the nematode *Ascaris*, which seem to require access to some oxygen (Nicholas 1975). *Ascaris* is capable of collagen synthesis, which also requires oxygen (Fujimoto and Prockop 1969).

A parasitic mode of life would seem to be essential for the evolution of large anaerobic metazoa, which have almost unlimited access to organic substrates and certain biomolecules which are synthesized by the aerobic host or derived from its food. On the other hand, this particular habitat does allow for at least periodic access to microaerobic conditions and the parasitic helminths have not totally severed the adaptations to the oxic world enjoyed by their free-living ancestors.

3.5.2 Anaerobic meiofauna?

The meiofauna is a collective term for aquatic metazoa which measure from about 0.1 to a few mm, a size range which overlaps with the larger protozoa. The most diverse meiofauna is found in the interstitia of marine sands and includes not only groups which are always characterized by small size (turbellarians, nematodes, rotifers, gnathostomulids, gastrotrichs), but also small representatives of virtually all other invertebrate phyla (Fenchel 1978; Swedmark 1964). The great majority of these forms are aerobes which occur only in the superficial, oxygen-containing sediment layers or in coarse sands with some advective water flow. Other forms, however, seem to occur exclusively in the anaerobic strata of sediments although only very few are found in the sulphidic zone. The question of whether these forms are true anaerobes, or how they might otherwise cope, has been debated for the past 25–30 years. A persistent difficulty has been their size: meiofauna are too small to study metabolic pathways or to establish the presence of particular enzymes in single individuals and it has not yet been possible to grow the assumed anaerobic or microaerobic forms in mass cultures in the laboratory. The unambiguous demonstration of the existence of obligate anaerobic protozoa from similar habitats quickly followed the establishment of laboratory cultures.

The ability of some aquatic meiofauna to survive extended periods of anoxia is well documented. Thus the benthic fauna in periodically stratified lakes survive anoxia during stratification episodes. Por and Masry (1968) studied the benthic meiofauna of Lake Tiberias, Israel. This lake remains stratified and the hypolimnion anoxic, for 8–9 months of each year. It was found that the nematode *Eudorylaimus andrassy* and the oligochaete *Euilodrilus heussleri*, in addition to two unidentified species (the larvae of a chironomid and a rhabdocoel turbellarian) survived for the duration of this period of anoxia. The larvae of *Chironomus anthracinus* are also known to survive long periods of anoxia, but growth is arrested until oxygen becomes available (Jónasson 1972). Several authors have demonstrated long-term survival of various small metazoa, especially certain nematode species under anaerobic conditions in the laboratory (e.g. Banage 1966; Wieser and Kanwisher 1961). Nematodes are frequently found in sulphidic and strongly reducing muds (e.g. Dye 1983). Zajcev *et al.* (1987, cited in Giere 1992) reported living nematodes in the sediments of the Black Sea at depths between 200 and 600 m; in this case it is difficult to imagine how these animals can ever get access to oxygen. Tubificid oligochaetes are also known to be very resistant to anoxia and to use fumarate reduction for their energy metabolism (Schöttler 1977). Famme and Knudsen (1984, 1985) grew a *Tubifex* sp. in an anaerobic microcosm and observed both growth and reproduction. This experiment could be a *bona fide* demonstration of the completion of a metazoan life cycle under strictly anaerobic conditions, but it has also been suggested that the experimental container did not, in fact, exclude traces of O_2 (Nicholas 1991) and attempts to reproduce the result have, unfortunately, not been made. Fenchel (1969, unpublished observations) found that a rotifer (*Colurella* sp.) and chaetonoid gastrotrichs sometimes increase in numbers in anaerobic, sulphidic incubations of marine sediments; this provides some evidence that a few small metazoa may, at least for a period, reproduce and grow under strictly anaerobic conditions.

The interstitial fauna of marine sediments have drawn special attention with respect to the relationship with oxygen. Most species from the majority of the groups represented, e.g. all crustaceans, are confined to the oxic layers and they are sensitive to anoxia and to sulphide. A number of species, however, occur at a depth in the sediments were oxygen is not detectable. As in the case of protozoa, it is possible to distinguish between three groups of meiofauna: those occurring exclusively in the oxic zone, those largely confined to the suboxic, but non-sulphidic zone and a few species which are found in the sulphidic zone. Compared to the protozoa, however, a much smaller fraction of the species occur in the anaerobic sediment (Boaden 1975; Boaden and Platt 1971; Fenchel 1969: Fenchel and Riedl 1970; Jensen 1983, 1986; Jensen *et al.* 1992; Ott and Schiemer 1973; Schiemer 1973; Wieser *et al.* 1974; Fig. 3.23). Similar faunas have been found in association with sulphide

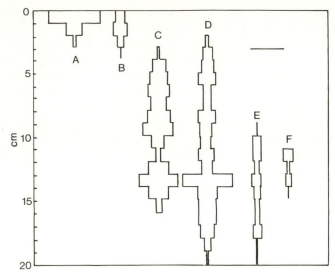

Fig. 3.23 Examples of the vertical distribution of nematodes in a sandy sediment at 15 m depth in the Sound, Denmark. A, *Monoposthia* sp.; B, *Viscosia franzii*; C, *Spirinia* sp.; D, *Leptomnemella aphanothecae*; E, *Odontophora rectangula*; F, *Daplonema* sp. Altogether the sample contained 88 species of nematodes. The sediment was sulphidic beneath 4 cm depth, but the sulphidic zone was penetrated by polychaete burrows. *Leptonomella* harbours ectosymbiotic sulphide-oxidizing bacteria, so that it must have some access to O_2. Scale bar, 2.5 individuals/cm² (data from Jensen 1987*b*).

seeps on the sea bottom (Jensen 1986; Powell *et al.* 1983). Those species which are found beneath the oxic zone are mainly representatives of the nematodes, gnathostomulids, gastrotrichs and solenofilomorphid and catenulid turbellarians.

Some authors (Boaden 1975; Fenchel and Riedl 1970) have suggested that the fauna of the deeper parts of the sediment (for which Boaden coined the term 'thiobios' since they occur in or in the vicinity of the sulphidic zone) consists of anaerobes and, since it is largely composed of primitive metazoa, that the fauna represents some sort of relict from a pre-oxic era. This latter idea is certainly wrong; the metazoa and, in fact most unicellular eukaryotes arose and diverged under aerobic conditions (see Sections 1.2, 3.1 and 5.3). Conversely, Reise and Ax (1979) claimed that such an anaerobic meiofauna does not exist, although some of these organisms do occur at low oxygen tensions or, periodically, under anoxic conditions.

Attempts to establish the precise relationship to oxygen of the 'thiobios' have so far proved elusive or the evidence is at best circumstantial. All species tested are capable of oxygen uptake (e.g. Ott and Schiemer 1973; Schiemer 1973), but so are anaerobic protozoa (Section 3.3). Some of the nematode species which occur in anaerobic sediments are sensitive to or even killed by exposure to atmospheric p_{O_2} (Wieser *et al.* 1974; P. Jensen, personal communication), but so are many microaerophilic protozoa (Section 3.3).

The truth is probably that the majority of this fauna consists of microaerophiles, which like many other aquatic invertebrates tolerate anoxia for extended periods of time. The small size of these organisms means that they can function as aerobes at oxygen tensions around or below the detection limit (Powell 1989; see also Section 5.3 and Fig. 5.9). Most of this fauna is found above the sulphidic zone in the suboxic zone. This layer may well be heterogeneous and contain microaerobic niches. The surroundings of polychaete and bivalve burrows represent such niches and, in fact, many meiofauna species characteristic of the suboxic strata concentrate around such burrows (Reise 1981*a,b*, 1984). The relatively high density of mitochondria in these forms compared to the 'oxybiotic' meiofauna (Duffy and Tyler 1984) also suggests that they are adapted to respiration at very low oxygen tensions. This is also supported by the fact that the thiobiotic meiofauna consist of species which are smaller and more slender than their oxybiotic relatives. Nematodes and oligochaetes which harbour sulphide-oxidizing bacteria (e.g. Ott *et al.* 1982; see also Section 5.1) occur in this zone and they must depend on microaerobic conditions. Many ciliates which occur in the suboxic zone above the sulphidic layers (e.g. loxodids) are microaerophiles. They apparently survive anoxia indefinitely, but they are capable of oxidative phosphorylation, and in oxygen gradients they accumulate at oxygen tensions of 2–5% atm. sat. (see also Sections 3.2 and 5.1). Some meiofauna species have been shown to perform vertical migrations in the sediment (Boaden and Platt 1971), so they may spend periods in the anaerobic zone. A few species, mainly nematodes, do apparently occur almost exclusively at considerable depth in the sulphidic zone where they make a living as grazers of bacteria. These worms remain potential candidates for being true metazoan anaerobes, but only laboratory cultures can establish this as a fact.

3.5.3 *Sulphide toxicity and detoxification*

The production of hydrogen sulphide through dissimilatory sulphate reduction is the most important process in anaerobic marine sediments and anoxic water columns, and concentrations in the sulphidic zone may exceed 10 mm. High concentrations (up to 3 mM) may also occur in some limnic systems (see Finlay *et al.* 1991). Most of this toxic compound is re-oxidized in the upper layers of sediments, but contact with sulphide is unavoidable for sediment-dwelling animals and occasionally sulphide also finds its way to the water column. Sulphide is very soluble in water. It occurs in the equilibrium: $H_2S \rightleftharpoons HS^- + H^+ \rightleftharpoons S^{2-} + 2H^+$, with $pK_1 \approx 7$ and $pK_2 \approx 13$

(Millero 1986), so that at pH values typical for natural waters most is in the form of HS⁻. Since the gaseous form enters cells more easily than the ionized forms, the toxicity increases with decreasing pH.

The most important toxic principle of sulphide is the binding to cytochrome c oxidase; in this way it acts as a respiratory inhibitor similar to CN⁻ (cyanide). It also binds to haemoglobin. Sulphide is therefore extremely toxic to aerobic organisms and exposure to micromolar concentrations is lethal for species without special detoxification mechanisms. Sulphide toxicity also plays a role for sulphide-oxidizing bacteria, which tolerate only limited concentrations. At higher concentrations, sulphide is also toxic to anaerobes such as phototrophic sulphur bacteria and sulphate reducers (e.g. Veldkamp and Jannasch 1972). Anaerobic eukaryotes are surprisingly tolerant and can grow at sulphide concentrations up to about 10 mM. The mechanisms of sulphide toxicity in anaerobes are not well understood, but they probably involve the inhibition of other metallo-enzymes and the reduction of protein disulphide bridges (Vetter *et al.* 1991).

Aerobic organisms show considerable variation with respect to sulphide tolerance, and species which are likely to receive transient or constant exposure in their natural habitats, are surprisingly resistant. An example is shown in Table 3.3 for three species of ciliates. In this case, the mechanism is not known, but since sulphide tolerance parallels tolerance to anoxia and to cyanide it may simply be due to the differential ability to survive on the basis of anaerobic metabolism. It has recently been shown that a variety of detoxification mechanisms occur in different animals; the topic has been reviewed by Vetter *et al.* (1991) and by Vismann (1991*a*).

Burrowing invertebrates are partly protected by the diffusion of oxygen from their irrigated tubes to the surrounding sediment (Section 4.1): a zone which is colonized by sulphide-oxidizing bacteria. The colonization of sulphide oxidizers on the surface of certain species of worms, or the presence of symbiotic sulphide oxidizers in gills or in the peripheral part of the body, also protects the host from sulphide poisoning (see Section 5.1). Some species seem to be capable of precipitating ferrous sulphide on their body surface, thus inhibiting the flow of dissolved sulphide into the body. A variety of biochemical detoxification mechanisms has been found in different species, including sulphide-binding proteins in the blood (Arp *et al.* 1984) and various enzymatically catalysed oxidations producing harmless sulphur species $S_2O_3^{2-}$, S^0 or SO_4^{2-} (e.g. Nuss 1984; Powell *et al.* 1979, 1980; Vismann 1990, 1991*b*). It is particularly interesting that some species have mitochondrial sulphide oxidation, which may be coupled to ATP-generation (Bagarinao and Vetter 1990). The gutless clam *Solemya* depends on symbiotic, chemoautotrophic sulphur bacteria. Sulphide oxidation to thiosulphate takes place in the mitochondria of the bivalve; the $S_2O_3^{2-}$ is then transported with the haemolymph to the symbionts and oxidized to sulphate (Powell and Somero 1986). The amoeba *Acanthamoeba* also oxidizes sulphide; low concentrations stimulate oxygen uptake and the oxidation is also likely to be mitochondrial (Lloyd *et al.* 1981). There is no evidence to show that more sulphide-tolerant cytochrome oxidases occur in these species (Vetter *et al.* 1991). A last resort for some organisms such as the priapulid worm *Halicryptus* is to rely on anaerobic metabolism when the sulphide flux into the animal exceeds the capacity to detoxify it (Oeschger and Vetter 1992).

Sulphide production in anaerobic habitats is one property which strongly affects the surrounding aerobic communities. Other such interactions between oxic and anoxic communities are discussed in Sections 5.1 and 5.2.

4

The structure of anaerobic communities

To animal and plant ecologists the term community structure means the description of ecological systems in terms of interactions (such as prey–predator relations, competition and mutualism) between species populations. Regarding microbial communities, the basic properties of syntrophic interactions and the competition for substrates between pairs of species has already been discussed in Sections 2.1–2.3 and 3.4. In microbial ecology, an additional approach has been to consider functional groups (rather than individual species) as the basic units of the communities. Prokaryote communities are thus studied in terms of syntrophic and competitive interactions between different physiological types of organisms (fermenters, sulphate reducers, etc.). The demonstration of particular types of energy metabolism (many of which have been discovered within the last two decades) and of other physiological properties and constraints in pure cultures remain of paramount importance for guiding our understanding of microbial communities. But the underlying principle of the approach is the thermodynamics of different types of energy metabolism, and the boundary conditions for the community structure are given by physical properties such as diffusion coefficients, rates of sedimentation, the penetration of light and the turn-over in 'flow-through' systems such as the rumen. This biogeochemical approach has been especially successful due to the development of sensitive tools such as microelectrodes (Revsbech and Jørgensen 1986) and gas chromatography, the use of radioactive or stable isotopes as tracers (e.g. Jørgensen 1978a,b; Oren and Blackburn 1979), and the use of more-or-less specific inhibitors for particular microbial processes (Oremland and Capone 1987). In this way it has been possible to quantify chemical gradients and quantify process rates with a high spatial and temporal resolution.

The first section in this chapter will be devoted to this approach. We will thereafter consider eukaryotic biota and the role of phagotrophy (predation) in anaerobic communities in some detail.

4.1 Spatial and temporal distributions of organisms and processes

4.1.1 *The redox-sequence in time and space*

In Sections 1.1 and 1.2 and Fig. 1.2 it was shown that in the heterotrophic mineralization of organic substrates, the electron acceptor of choice is that which yields the highest free energy in the process. The principle electron acceptors for respiration or methanogenesis are O_2, NO_3^-, SO_4^{2-} and CO_2 listed in order of decreasing capacity as oxidants, although compounds of intermediate oxidation level (notably N_2O, NO_2^-, $S_2O_3^{2-}$ and S^0) also play a role. The sulphate reducers and the methanogens are obligate anaerobes with a relatively restricted repertoire regarding potential substrates. Thus, they depend on coexisting fermenting bacteria for the degradation of carbohydrates and other organic compounds to low molecular weight fatty acids and hydrogen. Nitrate reducers are mainly facultative anaerobes which can degrade a wide variety of substrates (Section 2.2).

Chemical species other than the above-mentioned respiratory electron acceptors are also reduced in anaerobic waters. Thus oxidized manganese and iron are reduced and appear as dissolved Mn^{2+} and Fe^{2+} in anaerobic water (in sulphidic habitats, however, iron is precipitated as FeS and pyrite). It has been suggested that these metals also function as electron acceptors in electron transport phosphorylation (e.g. Roden and Lovely 1993). There is no doubt that the reduction of Fe and Mn is catalysed by microbial activity. Mechanisms include the spontaneous reduction by metabolic end-products (volatile fatty acids, H_2) and enzymatic catalysis wich is not coupled to energy conservation. Jones *et al.* (1983, 1984) and Sørensen (1982) found evidence that in the absence of nitrate, nitrate-reducing bacteria catalyse the reduction of Fe using the enzyme nitrate reductase. However, there is still no strong evidence that microbially mediated Fe- and Mn-reduction is coupled to energy conservation in a respiratory process (for reviews see Ghiorse 1988, 1989; Howarth 1993 and Section 2.2). The absence of free S^{2-} in the upper part of the anaerobic zone in the presence of active sulphate reduction is due to abiological oxidation of sulphide by oxidized Mn and Fe.

In a microbial succession based on the accumulation of organic material we expect a sequential depletion of O_2, NO_3^- and SO_4^{2-}, and eventually methanogenesis will prevail as the terminal mineralization process. In sediments, a similar vertical zonation will be found: at the surface, aerobic respiration takes place. At some depth where O_2 is depleted, a zone of nitrate respiration will be found. Below this, a zone of sulphate reduction will occur and finally, when SO_4^{2-} has been depleted, methanogenesis will take place.

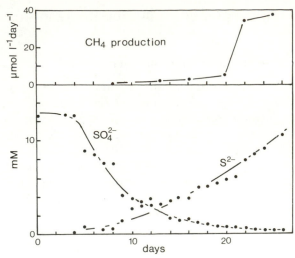

Fig. 4.1 Development of a $10\,\ell$ system with brackish water, dried lettuce leaves (corresponding to 0.4 g/ℓ), 5 ml estuarine sediment and incubated under a N_2-atmosphere. The sum of S^{2-} and SO_4^{2-} is not constant due to a transient accumulation of other sulphur compounds. Methanogenesis is insignificant until the SO_4^{2-}-concentration is < 1 mM.

That this may be a realistic description is exemplified in Figs 4.1–4.5. The reason why nitrate shows a maximum concentration at the oxic–anoxic boundary is that in the presence of oxygen, nitrate is produced by nitrifying bacteria from ammonia which diffuses upwards from the anaerobic zone. In the photic zone of the water column and at the sediment surface, NO_3^- is assimilated by photosynthetic organisms. The nitrate produced in the lower part of the oxic zone therefore diffuses upwards towards the photic zone as well as downwards to the anoxic zone where it serves for denitrification.

While the successional or spatial zonation of these processes is explained by thermodynamic considerations, the quantitative importance of the different electron acceptors is determined by two other factors. The first of these is the availability of the different electron acceptors. At atmospheric pressure, water contains about 250 µM of O_2 (depending on salinity and temperature) while seawater contains about 25 mM SO_4^{2-}. One mole of oxygen can oxidize one mole of [CH_2O] while one mole of sulphate can oxidize two moles of [CH_2O]. Thus, air-saturated seawater contains about 200 times more oxidation equivalents in the form of sulphate than in the form of oxygen. In freshwater, the sulphate concentration is variable, but much lower (typically 0.03–0.3 mM in eutrophic lakes). Sulphate is almost a conservative constituent of aerobic water (algae and bacteria do take up sulphate for assimilative reduction, but the effect on the concentration is negligible). In contrast, NO_3^- is consumed, not only by denitrification under

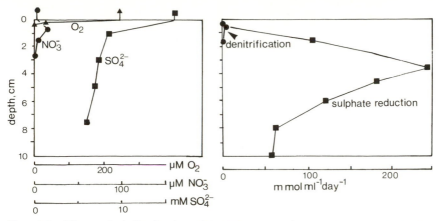

Fig. 4.2 The vertical distribution of O_2, NO_3^- and SO_4^{2-} and of rates of denitrification and sulphate reduction in an estuarine sediment (data from Sørensen *et al.* 1979).

anaerobic and microaerobic conditions, but also assimilated by aerobic phototrophs and by heterotrophs. Reactive N is often a limiting factor for primary production in aquatic habitats and nitrate usually occurs in very low concentrations. Denitrification therefore almost always plays a relatively small quantitative role in terms of C-oxidation in aquatic sediments (cf. Fig. 4.2). In sediments of temperate waters, the relative role of denitrification is somewhat higher during winter, when the demand of phototrophic organisms is lower and when the oxic surface zone is thicker. This allows for a more efficient nitrification of ammonia which otherwise tends to escape to the water column prior to its oxidation (Fenchel and Blackburn 1979; Sørensen *et al.* 1979; Vanderborght and Billen 1975).

The other factor which determines the relative role of electron acceptors is the availability of organic substrates at the different depth zones or stages in a succession. In an idealized sediment, organic material is provided only by sedimentation from the water column and the bulk of this material will be mineralized at or very close to the surface of the sediment. Since sedimentation is a relatively slow process, little degradable organic material will occur at greater depths in the sediment. Assuming that the degradation of organic material is a first order process (Berner 1971), heterotrophic activity will decrease exponentially with depth and only organic materials which are resistant to degradation will accumulate. In real sediments, the upper *ca* 10 cm are typically mixed due to animal activity (bioturbation), and in most shallow water sediments the surface layers are periodically eroded by currents and wave action. At other times material is re-deposited. In this way, organic material is more or less regularly buried in the upper layers of the sediment, thus sustaining microbial activities at greater depths.

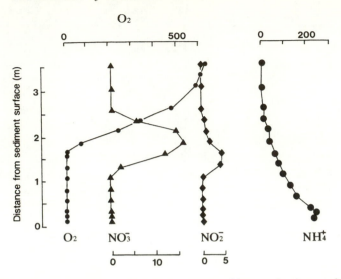

Fig. 4.3 Vertical profiles of dissolved oxygen and inorganic nitrogen in a stratified freshwater pond. As oxygen becomes depleted, nitrate is used as alternative electron acceptor, being reduced to nitrite which accumulates close to the top of the anoxic zone. Nitrate is continuously replenished in the oxic zone by nitrifying organisms oxidizing the upwards diffusion of ammonia. All concentrations are µM (redrawn from Finlay 1985).

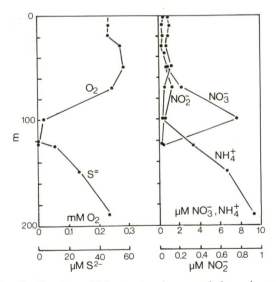

Fig. 4.4 The distribution of N-compounds around the oxic–anoxic boundary in the Fårö Deep in the Baltic Sea. Nitrite is an intermediate product of nitrification of ammonia (data from Rheinheimer *et al.* 1989).

Even so, heterotrophic activity decreases with depth and anerobic processes and anaerobic microbial biomass are usually highest immediately beneath the oxic zone. Thus, Jørgensen (1977b) found that beneath a depth of about 10 cm, sulphate reduction rate decreased rapidly with depth, reflecting a decreasing amount of degradable compounds (Fig. 4.16). As a result of the decrease in concentration and degradability of organic substrates with depth, the electron acceptors which are utilized first tend to play a relatively larger role in mineralization than their relative abundance would indicate. An increasing import of organic material will mean that an increasing amount is mineralized via sulphate reduction and methanogenesis relative to aerobic respiration. In shallow, productive marine sediments, sulphate reduction may account for more than 50% of the terminal mineralization, but in less productive offshore sediments, this figure is reduced to 10–20% (Jørgensen 1977b; Jørgensen and Revsbech 1989).

4.1.2 *Methanogenic versus sulphate-dependent systems*

The quantitatively most important anaerobic terminal mineralization processes are sulphate reduction and methanogenesis. Even by the last century, it was demonstrated that when cellulose is degraded anaerobically in the presence of gypsum, the reduction equivalents of the organic substrate appear as H_2S, but in the absence of sulphate, CH_4 is produced instead (Widdel 1988). It has also long been known that in the presence of active sulphate reduction, there is very little or no methanogenesis. It has been suggested that methanogens are inhibited by either sulphate or by sulphide, but this has been shown not to be the case. It is now established that the reason is solely competition for the two substrates common to methanogens and sulphate reducers: acetate and H_2. Studies on the growth kinetics with these substrates in pure cultures of different species of sulphate reducers and of methanogens show that sulphate reducers are superior competitors in accordance with the thermodynamics of the processes (Ward and Winfrey 1985; Widdel 1988). The best measure of the competiveness at low substrate concentration is the ratio μ_m/K_s, where μ_m is the maximum growth rate (at high substrate concentrations) and K_s is the half-saturation constant in the Monod-equation. The ratio represents the slope of the Monod-equation at low substrate concentrations, where growth rate is almost linearly proportional to concentration and sulphate reducers have higher values of μ_m/K_s than do methanogens for competitive substrates. Sulphate reducers can grow at lower substrate concentrations than can methanogens. The hydrogen pressure is therefore lower in the presence of sulphate reduction (e.g. in marine sediments) than in methanogenic systems (e.g. sulphate-depleted lake sediments, the rumen).

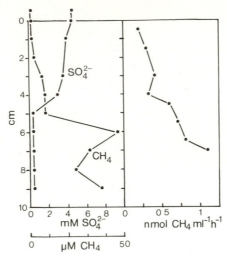

Fig. 4.5 The vertical distribution of sulphate, methane and the rate of methano-genesis in a sediment from Nivå Bay, May 1992 (data from Fenchel 1993).

The ability of sulphate reducers to compete will, of course, at some point decline with decreasing sulphate concentration. However, as with aerobic microorganisms, they are unaffected until the electron acceptor concentration is very low. Thus, Ingvorsen and Jørgensen (1984) found half-saturation concentrations within the range 4.8–77 µM SO_4^{2-} for different species of sulphate reducers. However, a significant increase in the rate of methano-genesis can be observed in marine sediments when the sulphate concentration is below ≈ 1 mM.

Due to the high concentration of sulphate in seawater, sulphate reduction is always the dominating process in marine anoxic habitats, typically accounting for 90–99% of anaerobic terminal mineralization (Ward and Winfrey 1985). Only in productive areas will sulphate be totally depleted at a depth in the sediment where there is still an appreciable amount of degradable organic material (compare Figs 4.5 and 4.16). Even so, there is always a low rate of methane production in sulphate-containing interstitial water (e.g. Senior *et al.* 1982; Fig. 4.5). One reason for this is the existence of non-competitive substrates for methanogens (Oremland and Polcin 1982). Thus methanol and methylamine are substrates for methanogens, but not for sulphate reducers. The addition of these compounds to incubated samples of marine sediments stimulates methanogenesis. When methano-genesis is specifically inhibited by adding BES (2-bromoethanesulfonic acid; see Oremland and Capone 1987) to sediments, methanol and methy-lamine accumulate. Another explanation for some methanogenic activity in sulphate-containing sediments may be the existence of sulphate-free microniches inside detrital particles. Finally, in some habitats (sulphureta, sandy sediments), protozoa with endosymbiotic methanogens (which are protected from competition with sulphate reducers) may contribute measurably to the methane production (Fenchel 1993; see also Sections 3.4 and 4.2).

The importance of sulphate (and S^0) as terminal electron acceptors in marine anaerobic habitats is immense. The complete mineralization of 1 kg of organic material yields 570 g H_2S, so copious amounts of this toxic compound are produced in the sea bottom. However, as sulphide diffuses upwards, it is almost completely oxidized by chemolithotrophic bacteria in the lower part of the oxic zone (5.1), or it may be oxidized anaerobically by phototrophic sulphur bacteria in the presence of light (4.3). Smaller amounts are deposited as a type of fossil fuel in the form of iron sulphides, and only trace amounts escape to the oxic part of the water column or into the atmosphere.

The substrates of sulphate reducers in a marine sediment were studied by Sørensen *et al.* (1981). These substrates (volatile fatty acids, H_2) normally occur in very low concentrations in interstitial water. By the (assumed) selective inhibition of sulphate reducers by MO_4^{2-}, the increase in substrate concentration could be followed (molybdate is a competive inhibitor of sulphate reduction, and must therefore be added in concentrations comparable to ambient SO_4^{2-}-concentration; see Oremland and Capone 1987). From the rates of linear increase in substrate concentrations, it was found that acetate constituted 40–50% of the substrates for sulphate reduction, H_2 constituted about 10% and propionate, iso-butyrate and butyrate were responsible for the rest. The authors also found a linear increase in CH_4 immediately after the molybdate-addition and interpreted this as a result of the relief from competition with sulphate reducers. However, it is more likely that methanogenesis was unaffected, but that the molybdate addition inhibited methane-oxidation by sulphate reducers (see below).

Although lake water contains much lower concentrations of SO_4^{2-}, sulphate reduction may still be an important process in the anaerobic water column and especially in the surface layer of the sediment (Ingvorsen *et al.* 1981; Ward and Winfrey 1985). Ward and Winfrey (1985) listed a number of eutrophic lakes in which between 20 and 70% of the terminal anaerobic mineralization is due to sulphate reduction. When the water column above the sediment is aerobic, there is a thin (1–3 cm) surface layer of sediment with very intense sulphate reduction; below this layer, sulphate is undetectable. The sulphide produced is rapidly re-oxidized at the surface of the sediment and re-cycled (with a turnover time of 1–3 h), and the sulphate concentration in the oxic surface of the sediment may exceed that of the water. In this way, a high rate of sulphate reduction can be maintained in spite of a low bulk concentration of SO_4^{2-} (Bak and Pfennig 1991). Westermann (1993) has recently compiled data on the relative importance of methanogenesis and sulphate reduction in wetlands and water-logged soils.

The substrates of methanogens in freshwater and marine sediments have drawn some interest. In addition to some quantitatively less important substrates (methanol, methylamine, dimethylsulphide) methanogens utilize H_2, acetate and formate; hydrogen and acetate are believed to be the most important (although the quantitative importance of formate in natural systems has not been studied). In contrast to the sulphate reducers, the methanogens cannot utilize higher fatty acids (e.g. butyrate and propionate) which in methanogenic systems must first be converted into acetate by proton-reducing acetogenic bacteria (see Section 2.1). The relative role of acetate and H_2 can be determined by incubating sediment samples with $^{14}CO_2$ or with ^{14}C-labelled acetate. If only H_2 is used, the activity of the labelled CO_2 will be found in the CH_4 which is produced, while the methane will remain unlabelled if only acetate is the substrate. If labelled acetate is used, the converse is true. In this way Cappenberg and Prins (1974) found that about 70% of the methane production in a lake sediment derived from acetate, but in other systems H_2 seems to play a relatively larger role. Since methanogens grow more efficiently with H_2 than acetate, however, the incubation of anaerobic sediment or water with H_2 always stimulates methanogenesis. Methanogens are always hydrogen-limited and maintain a low p_{H_2}, thus stimulating hydrogen-producing fermentation (Abram and Nedwell 1978; Oremland 1988; Oremland and Taylor 1978).

The methane produced in the anaerobic water column or in sediments is oxidized by methane-oxidizing bacteria in the oxic part of the water column or sediment. However, since the solubility of methane is low, much of the CH_4 may escape to the atmosphere by ebullition; a phenomenon which is easily observed in marshes and swamps. It has, however, been a controversial question whether CH_4 can be oxidized anaerobically by sulphate reducers. Methane profiles in marine sediments often show a concave-upward shape in the sulphate-containing but anoxic zone. Assuming a steady state situation, this can only mean that methane is consumed. Examples are the methane-profiles published for sediments of the Long Island Sound by Martens and Berner (1974); see also Fig. 4.5. The potential gain in free energy from methane oxidation with sulphate is low (cf. Fig. 1.2) and it has not been possible to obtain pure cultures of sulphate reducers which can grow with only CH_4 as substrate. However, there is convincing evidence to show that anaerobic oxidation of CH_4 does take place in marine sediments and that it is probably slowly co-metabolized with other substrates by sulphate reducers (Alperin and Reeburgh 1985; Iversen and Blackburn 1981; Reeburgh 1980).

Systems in which methanogenesis is the dominant terminal mineralization step include the microbial biota of intestinal tracts such as the rumen

(Hungate 1966), the termite hindgut (Odelson and Breznak 1983) and the intestinal tract of other animals including man (Oremland 1988). In these cases, the availability of oxidized sulphur compounds is very low. It is more intriguing how marine animals with fermenting bacteria in the intestinal tract (green turtles; see Fenchel *et al.* 1979; manatees and dugongs, sea-urchins) apparently inhibit sulphate reduction and so avoid the accumulation of toxic sulphide. No attempts have been made to answer this question.

Methane concentrations in excess of what would be in equilibrium with the atmosphere have, on several occasions, been found at certain depths in the oceanic water column (Scranton and Brewer 1977; Oremland 1988). Since methanogenesis is an exclusively anaerobic process, the explanation for this phenomenon has been debated. Oremland (1988) suggested that the methane derives from the intestines of planktonic organisms and of fish and baleen whales (which, it has been suggested, have pregastric fermentation for the degradation of chitin: see Herwig and Staley 1986). An alternative explanation is methanogenesis in the anaerobic interior of marine snow (Bianchi *et al.* 1992).

4.1.3 *Anaerobic communities in the light*

The most complex and interesting anaerobic communities are those exposed to light. They are characterized by the presence of phototrophic bacteria, which are often so abundant as to become macroscopically conspicuous. These organisms have relatively large and characteristic shapes and colour and they can often be identified microscopically. The communities they form together with various heterotrophic and chemolithotrophic forms are characterized by steep chemical gradients and zonation patterns. Some of these communities resemble or are identical to the oldest known biotic communities on Earth (Section 1.2). In microbial mats in particular, the oxic and anoxic parts are so integrated that they must be treated together in a comprehensive description of these communities.

The physiology and energy metabolism of phototrophic bacteria was discussed in detail in Section 2.2. However, it will be useful to summarize briefly those properties which are necessary for understanding zonation patterns and other features of the integrated communities.

There are several groups of phototrophic bacteria. The cyanobacteria have chlorophyll *a*, and accessory pigments (phycobilins) which allow them to absorb light in the green part of the spectrum. They possess both photosystems I + II. Their photosynthesis is typically oxygenic and they are fundamentally aerobic organisms. However, in the presence of higher concentrations of sulphide, photosystem II is inhibited and many species are capable of anoxygenic photosynthesis on the basis of photosystem I and with sulphide as electron donor. At least in some systems, they have been shown to play an important role in anaerobic sulphide oxidation (for a review see Cohen *et al.* 1986).

The principal photosynthetic pigments of the purple sulphur bacteria are bacteriochlorophyll *a* (with an infrared absorption maximum within the range of 825–890 nm) and bacteriochlorophyll *b* (absorption maximum 1020–1040 nm). The purple sulphur bacteria utilize S^{2-} or S^0 as electron donors and produce S^0 or SO_4^{2-} as end-products in the light; in the dark their survival is based on heterotrophy under anaerobic or microaerobic conditions. The purple sulphur bacteria often occur in conspicuous masses. The purple non-sulphur bacteria also have either bacteriochlorophyll *a* or *b* as the dominant photosynthetic pigment. They are metabolically versatile: in the light they are anaerobes and perform anoxygenic photosynthesis with H_2, various low molecular weight organic compounds (e.g. acetate) or in some cases S^{2-} or Fe^{2+} as electron donors. In the dark they are microaerobic respirers. The purple non-sulphur bacteria do not show mass occurrence in nature, but they are omnipresent in all the communities described below. Many representatives of the purple bacteria are motile and show a pronounced chemosensory and photosensory behaviour.

The green sulphur bacteria are strict anaerobes; they use S^{2-} as an electron donor and they produce elemental sulphur as a metabolic end-product. Green sulphur bacteria have bacteriochlorophyll *c* (infrared absorption peak: 745–755 nm) or *d* (705–740 nm). In nature, green sulphur bacteria live in close syntrophic relationship with sulphur-reducing heterotrophs. Self-sustaining (requiring only light) closed co-cultures of green sulphur bacteria and sulphur-respiring heterotrophs can be maintained in the laboratory. '*Chlorochromatium*' and '*Pelochromatium*' are symbiotic associations between a green sulphur bacterium and a heterotrophic sulphur reducer.

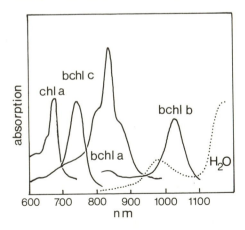

Fig. 4.6 The absorption spectra of chlorophyll *a* and the three most important bacteriochlorophylls in the red–infrared region (compiled from various sources).

Chloroflexus species have been discovered relatively recently in microbial mats in hot springs, but microscopical evidence suggests that they are present in most microbial mats. They are green, filamentous gliding organisms with bacteriochlorophyll similar to bacteriochlorophyll *c*, but they are photoheterotrophs and not related to the green sulphur bacteria. For the sake of completeness, the group heliobacteria have recently been discovered in soils; these are strict anaerobic phototrophs, but their ecological role is still not well understood (Madigan 1988).

The characteristic colours of photosynthetic bacteria are not due to the bacteriochlorophylls (extracts of which are light blue or almost colourless), but to carotenoids, which act as accessory photosynthetic pigments in the blue region of the spectrum. As with the different bacteriochlorophylls, the carotenoids in different species differ in their absorption spectra.

The factors which determine coexistence and zonation patterns in nature are the differential light absorption spectra (Fig. 4.6), and differential tolerance and requirements for sulphide and oxygen (Madigan 1988; Pfennig 1989; Van Gemerden 1974, 1993; Van Gemergen and Beeftink 1983; Van Gemerden and de Witt 1989). In most cases anoxygenic phototrophs live beneath a layer of oxygenic phototrophs (phytoplankton in lakes, cyanobacteria and eukaryotic algae in sediments and in stromatolitic microbial mats) and so it is clearly adaptive that the long-wave absorption peaks of bacteriochlorophylls are displaced relative to the red absorption peak of chlorophyll *a*. In mineral sediments, the increasing diffraction and scattering with decreasing wavelength means that long wavelength light penetrates deeper; the attenuation of blue light is about ten times greater than that of infrared light at a depth of 4 mm (Fenchel and Straarup 1971; Fig. 4.7). More recently, it has been possible to study the attenuation of different parts of the light spectrum of microbial mats *in situ* using fibre-optic microprobes

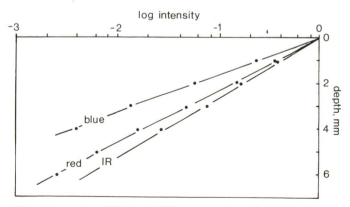

Fig. 4.7 The attenuation of light of different wavelengths in clean quartz sand in seawater (data from Fenchel and Straarup 1971).

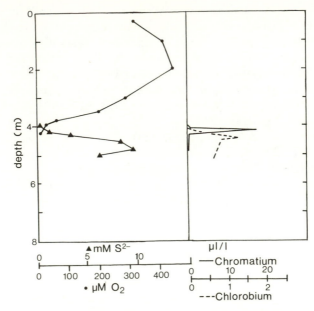

Fig. 4.8 The vertical distribution of O_2 and total S^{2-} and of the dominating phototrophic bacteria in Lake Vilar, Spain (redrawn from Guerrero *et al.* 1985).

(Jørgensen and Des Marais 1986*a*). In sediments, the blue part of the absorption spectrum plays a relatively small role for phototrophic bacteria. In the water column, however, blue light usually penetrates deeper than light of longer wavelengths.

In lakes with an anoxic water column which reaches the photic zone, a distinct layer of purple sulphur bacteria forms immediately below the oxycline. Below these purple bacteria there is typically a layer of green-sulphur bacteria. An example is shown in Fig. 4.8 from a Spanish meromictic lake. In this lake the purple bacteria are dominated by *Chromatium* and the green bacteria by *Chlorobium* (Guerrero *et al.* 1985). In other lakes, a more diverse assemblage of purple sulphur bacteria are found; thus in some stratified lakes in Michigan, Caldwell and Tiedje (1975*a,b*) recorded successive layers of *Thiopedia*, *Thiospirillum* and *Thiocystis*.

In accordance with the light-dependent oxidation of sulphide, the anoxic layer migrates upwards during the night, when only chemolithotrophic bacteria oxidize sulphide in the lower part of the oxic zone. In Solar Pond, Sinai, the chemocline migrates about 30 cm upwards during the night. In this unusual system, the cyanobacterium *Oscillatoria limneticum* is the most important phototrophic sulphide oxidizer, rather than purple bacteria (Jørgensen *et al.* 1979).

The layer of purple sulphur bacteria in lakes harbours many species of phagotrophic eukaryotes, several of which appear to depend on the phototrophic bacteria for food (Finlay *et al.* 1991; see also Section 4.2). Clarke *et al.* (1993) have recently described the life cycle of a prokaryote which is an ectosymbiont on the purple bacterium *Chromatium weissei.* The host reaches extremely high population densities close to the top of the sulphide-rich anoxic hypolimnion in some Spanish lakes. There is evidence that the *Chromatium* population secretes dissolved organic compounds, so attachment by the ectosymbiont is probably an adaptive feature allowing it to remain very close to the source of production of utilizable organic carbon. The development of a plate of phototrophic sulphur bacteria in permanently or periodically stratified lakes with an anoxic hypolimnion is a widespread phenomenon. Trüper and Genovese (1968) and Vicente *et al.* (1991) provide other examples.

Sheltered and shallow marine sediments often develop superficial microbial mats. These are not usually permanent structures; periodically they break up due to wave erosion. In temperate latitudes this is particularly evident during the winter, when the mats do not re-establish themselves as quickly as during summer. Microscopical observation and the extraction of photosynthetic pigments from core slices reveal the zonation of the mats (Fenchel and Straarup 1971; Nicholson *et al.* 1987; Pierson *et al.* 1987; see also Figs. 4.9 and 4.10).

In the Sippewissett salt marsh in Cape Cod, Massachusetts, the upper 1–2 mm (the 'golden layer') is dominated by diatoms and cyanobacteria (especially *Lyngbya* and *Phormidium*). Below this layer is found the 'upper green layer', dominated by the cyanobacteria *Oscillatoria* and *Phormidium.* At a depth of about 4 mm, the 'purple layer' is dominated by the purple sulphur bacteria *Thiocapsa, Chromatium* and *Amoebobacter.* Beneath this, a 'peach layer' harbours the purple sulphur bacterium *Thiocapsa pfennigii* (which has bacteriochlorophyll *b*). Finally, between this layer and the black, sulphide layer which starts at a depth of about 7 mm, there is a 'lower green layer' dominated by the green sulphur bacterium *Prosthecochloris.* The vertical distribution of chlorophylls is shown in Fig. 4.9 (Nicholson *et al.* 1987; Pierson *et al.* 1987).

The actual depth of the photosynthetic layer and its zonation pattern depend on the penetration of light, which again is partly governed by the optical properties of the mineral grains. In Nivå Bay, Denmark (Fenchel 1969; Fenchel and Straarup 1971; unpublished observations) the layer with photosynthetic organisms is only 5–5.5 mm thick (Fig 4.10). As in the Sippewissett mat, there is a yellow-green 1.5–2 mm thick surface layer with cyanobacteria and photosynthetic eukaryotes including diatoms, dinoflagellates, cryptomonads and euglenoids. Beneath this, a 1.5 mm thick layer with a dark blue-green colour is dominated by filamentous cyanobacteria

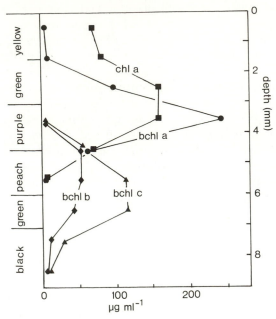

Fig. 4.9 The vertical distribution of different chlorophylls in a microbial mat from the Great Sippewissett salt marsh, Cape Cod, Massachusetts (redrawn from Pierson *et al.* 1987).

(*Phormidium*, *Oscillatoria*, etc.) which bind the sand grains to form a coherent and tough layer, in contrast to the more flocculent surface layer. Beneath this, there is a pink, 1.5–2 mm thick layer with purple sulphur bacteria (especially *Thiocapsa*, *Chromatium*, *Thiopedia* and *Lamprocystis*) and *Chloroflexus*-like filamentous bacteria. There is no macroscopically visible peach or lower green layers in this mat, so the pink layer rests directly on the black sulphidic sediment. However, green sulphur bacteria are numerous in the lower part of the mat. All layers harbour a diverse assemblage of unicellular eukaryotes (Fenchel 1969; see also section 4.2). Periodically during the summer, the mat flakes off the sediment surface and a new mat forms beneath it (Fig. 4.10).

Figure 1.9 shows the distribution of O_2 and of S^{2-} in this mat during darkness and after exposure to strong light for 2 hours. It demonstrates extreme changes in the chemical environment during the diurnal cycle. It also explains why most inhabitants of the mat are motile. Thus chemolithotrophic bacteria (*Beggiatoa*, *Thiovulum*; see Section 5.1) follow the oxygen–sulphide overlap zone during changes in light intensity. During calm summer nights they may even leave the sediment to form a white haze in the water, showing that the anoxic zone has risen above the sediment

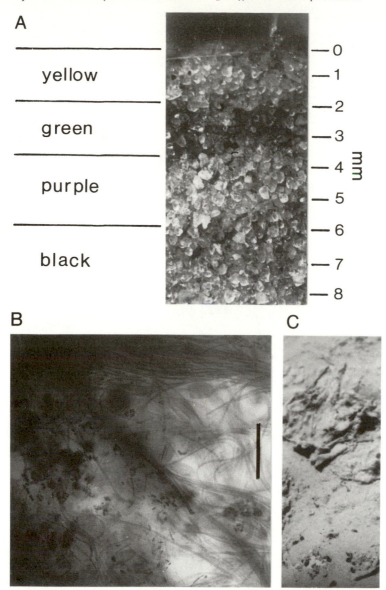

Fig. 4.10 Microbial mat from Nivå Bay in the summer. A, a core sample showing the different colour zones; B, the transition between the green and the purple zone showing filamentous cyanobacteria, colonies of purple sulphur bacteria attached to sand grains and a *Beggiatoa* filament. Scale bar, 100 μm. C, Macroscopic view of the mat at an occasion when waves had displaced some of the mat. It can be seen that the mat has mechanical coherence.

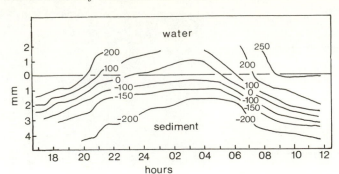

Fig. 4.11 The vertical redox profiles (mV) in a shallow sediment during summer in Limfjorden, Denmark from the afternoon to the following noon (redrawn from Hansen et al. 1978)

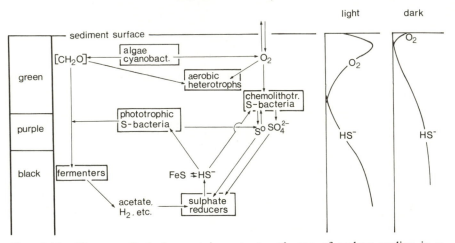

Fig. 4.12 The quantitatively most important pathways of carbon cycling in a microbial mat. The processes and the microorganisms migrate vertically according to the light intensity. The chemolithotrophic sulphide oxidizers track the vertical position of the oxygen–sulphide boundary (original).

surface (Fenchel 1969). Likewise, the purple *Chromatium* spp. follow the upper part of the sulphide zone. An example of the diurnal migrations of the chemical zonation is also shown in Fig. 4.11. Figure 4.12 shows the most important microbial processes and their zonation in a bacterial mat.

When such mats are covered by a very shallow layer of water, the entire water column may become anaerobic and sulphidic during calm nights with the result that some H_2S escapes to the atmosphere (Hansen *et al.* 1978). This effect is evident in the vicinitiy of salt marshes where the air often has a sulphidic smell only at night; during the day, the efficiency of phototrophic sulphur bacteria can be appreciated.

In some cyanobacterial mats, the oxic zone may extend deeper than the photic zone during the day. In such systems, anoxygenic phototrophs are excluded and sulphide oxidation is predominantly carried out by vertically migrating, chemolithotrophic bacteria (Jørgensen and Des Marais 1986a,b).

Banded bacterial mats are known from many coastal areas around the world (Cohen and Rosenberg 1989). Examples of other closely studied mats are those found in the Dutch Waddenzee (e.g. Visscher *et al.* 1992a,b).

In more extreme cases, where decaying seagrasses and algae accumulate in sheltered coves or lagoons, the sulphide production will be so high that the water column becomes anaerobic and sulphidic. Such shallow water 'sulphureta' (ecological systems in which the sulphur cycle dominates) appear intensely red or purple. This phenomenon regularly occurs during the summer in the innermost part of Nivå Bay where it is promoted by decaying *Zostera*, *Ulva* and *Enteromorpha*. The 10–20 cm deep water column is sulphidic at the surface even during the day and the S^{2-}-concentration may reach 10 mM a few cm below the surface, reaching even higher concentrations at the bottom. The water contains a dense suspension of purple sulphur bacteria (*Thiocapsa*, *Chromatium*, etc.) mixed with cyanobacteria (mainly *Oscillatoria*) and a great variety of heterotrophic bacteria and eukaryotic microorganisms (Fenchel 1969). Such 'red waters' caused by purple sulphur bacteria (as opposed to those in hyperhaline lagoons and salt evaporation basins which are caused by (aerobic) halophilic bacteria) have also been found and studied in lagoons along the Mediterranean Sea (Caumette 1986 and references therein).

Bacterial mats of hyperhaline and hyperthermal habitats (coastal hyperhaline lagoons, salt evaporation plants, hot springs) have drawn special attention. Their common characteristic is that they are undisturbed by bioturbation (metazoa are usually absent) and by hydrodynamic erosion. The mats can therefore grow continuously to form thick structures and they may be associated with calcium carbonate deposition. In this way they closely resemble Precambrian stromatolites (see Section 1.2). Mats of hot springs have been studied intensively in Yellow Stone Park, Iceland, New Zealand and elsewhere (see Brock 1978 and Ward *et al.* 1989), while the best studied hyperhaline microbial mats are those of lagoons in Baja California, Mexico, and in Solar Pond, Sinai Peninsula (D'Amelio *et al.* 1989).

Some of the hot spring mats are unusual; they may consist of only *Chloroflexus* and the cyanobacterium *Synechococcus* (Brock 1978) or be totally anaerobic, consisting of the green sulphur bacterium *Chlorobium* (Castenholz *et al.* 1990). The hyperhaline mats follow the general pattern described above for marine shallow-water mats: a superficial layer is dominated by diatoms and covers a thick layer of filamentous cyanobacteria lying over a layer of phototrophic sulphur bacteria.

The mats of Solar Pond and the vertical zonation of different species have been described in detail by Jørgensen *et al.* (1983). The cyanobacterial layer consists of several vertical zones with characteristic species assemblages. Excretions of polysaccharides by the cyanobacteria fill the interstitia between the filaments and give the mat a compact and tough structure. New growth continues at the top of the mat and results in annual laminations. The upper 4–6 mm of the mat is subject to extreme diurnal variations with respect to p_{O_2}, pH, and sulphide concentrations. During the day, 400% supersaturation with O_2, with consequent bubble formation, is found at a depth of 2 mm, but after several hours in the dark, this layer becomes completely anoxic. The sulphur cycle of the mats in Solar Pond was studied by Jørgensen and Cohen (1977). Close to the surface there was a very active sulphur cycle. Sulphate reduction could, however, be measured down to a depth of 80 cm in the mat, which on the basis of the laminations could be estimated to be about 700 years old. At this depth, the rate of sulphate reduction was only about 10^{-4} times that measured close to the surface, showing a continuous, but very slow degradation of the more refractile components formed by the photosynthetic activity at the surface.

4.1.4 *Temporal and spatial heterogeneity*

We have already discussed diurnal variation in vertical extension of the anaerobic and reducing zone in light-exposed anaerobic communities. Regular changes in the vertical extension of anaerobic conditions also occur on an annual basis. Figure 4.13 shows how reducing conditions approach the surface during summer and retreat again during winter in a lake sediment. Periods when the reducing zone is close to the surface reflect large fluxes of organic particles from the water column to the sediments. Such periods follow the decline of blooms of plankton algae and these occur several times between early spring and autumn. A similar pattern can be

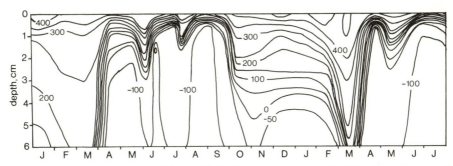

Fig. 4.13 The redox profiles (mV) during a two-year period in the sediment of Airthrey Loch, Scotland (redrawn from Finlay 1980).

observed in the sediments of coastal waters. In temperate waters there will typically be a spring diatom bloom and an autumn dinoflagellate bloom, and a substantial part of this production is not mineralized in the water column, but settles on the surface of the sediment.

Such patterns are also evident in shallower waters where the primary production (by benthic eukaryotic microorganisms, cyanobacteria and macroalgae) is highest during summer. If such sediments are exposed to waves, however, the pattern is more complex, because the sediment will periodically be stirred or eroded, resulting in the oxidation and infusion of SO_4^{2-} to a certain depth. Such events will also result in the burial of organic material and thus a subsequent development of anaerobic and reducing conditions (Fenchel 1969). In such cases, the sediment will be spatially heterogeneous and display different successional stages in the anaerobic degradation of organic material. The chemical gradients may also deviate from the previously described idealized picture. In particular, this applies to chemical species with a relatively low turnover rate. At the surface, the turnover time of oxygen is of the order of minutes or hours in productive sediments. A new steady state is therefore rapidly re-established and ideal gradients (e.g. Figs 1.1 and 1.9) are found and can be used for calculating fluxes. Gradients of sulphate, on the other hand, often deviate from the idealized situation and reflect past events such as local accumulations of organic material or recent infusions of sulphate due to tidal pumping or re-suspension of the upper part of the sediment. An anomalous vertical distribution of sulphate may also be found in estuarine sediments due to a varying salinity in conjunction with periodic infusion of fresh seawater into the sediment. This is because the large pool size of SO_4^{2-} leads to turnover times which must be measured in days or weeks.

An example of patterns resulting from strong wave action is shown in Fig. 4.14; the data are from a shallow beach in northern Jutland (Dando *et al.* 1993). A few days before sampling, a storm re-suspended the sediment to a depth of 10–15 cm, replenishing sulphate down to that depth. At the same time a ≈ 5 cm thick layer of decaying macroalgae was buried in the sediment. This layer already had a high intensity of sulphate reduction and methanogenesis (as revealed by the CH_4 and S^{2-} and SO_4^{2-} concentration profiles). These processes, however, could not quite cope with the activity of fermenters as demonstrated by the unusually high concentration of hydrogen, the low value of pH and a distinct smell of butyric acid. Such phenomena result in complex spatial and temporal patterns. They also secure renewal of organic material in the sediment and thus support higher rates of anaerobic metabolism and of biomass than is found at similar depths in more sheltered sediments.

The activity of burrowing animals is another factor which creates spatial heterogeneity. This occurs wherever the overlying water column is oxic. The

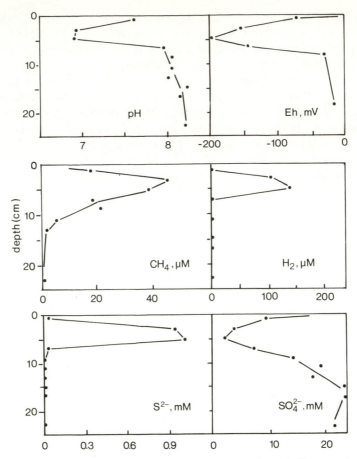

Fig. 4.14 Profiles of different chemical parameters in a shallow sandy sediment with a distinct layer of decaying macroalgae at a depth of around 5 cm (data from Dando *et al.* 1993).

activity of animals may simply consist of mixing the upper part of the sediment thus mixing particulate organic material into the sediment and by increasing the rate of sulphate flux over that possible by molecular diffusion. Many species (e.g. polychaetes, crustaceans and bivalves) build tubes which they irrigate with water from the surface. Most of this activity is confined to the upper ≈ 10 cm, but some animals can burrow much deeper, e.g. a species of the actinian *Cerianthus* which has been found to make burrows down to 40 cm in a reducing deep-sea sediment (Jensen 1992).

Animals which live in irrigated tubes produce surrounding oxidized zones, thus protecting themselves against sulphide (Fig. 4.15). Some invertebrates with symbiotic chemolithotrophic sulphur bacteria (such as species

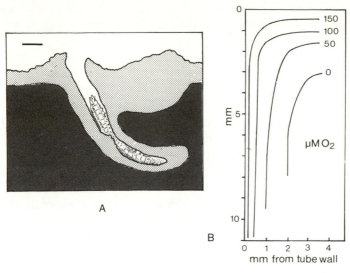

Fig. 4.15 A, a burrow of the polychaete *Nereis diversicolor* close to the wall of an aquarium. The dark sediment signifies the sulphidic zone; an oxidized zone surrounding the burrow is evident; scale bar: 5 mm (original). B, The oxygen concentrations around an irrigated, artificial tube placed in an anoxic sediment (redrawn from Meyers *et al.* 1988).

belonging to the bivalve genera *Myrtea* and *Thyasira*) use an extensive network of tubes to secure a supply of sulphide as well as of oxygen (Dando *et al.* 1985, 1986).

The surroundings of animal tubes form a habitat for chemolithotrophic bacteria and for microaerophilic protists and metazoa (Meyers *et al.* 1987, 1988; Reise 1981*a,b*, 1984). They also have a profound effect on the chemistry of the sediment by extending the surface area of the oxic–anoxic boundary and by importing oxidants (O_2, NO_3^-, SO_4^{2-}) to deeper layers in the sediment (Aller 1983; Aller and Yingst 1978, 1985; Hylleberg and Henriksen 1980).

Spatial heterogeneity on a much smaller scale is evident from the fact that opposed processes (e.g. denitrification and nitrification or sulphate reduction and sulphide oxidation) may take place within the same strata in a sediment and even within distances of < 1 mm. The spatial arrangement of syntrophic partners (such as the symbiosis between green sulphur bacteria and sulphur-reducing heterotrophs, see Section 2.3) represents heterogeneity on the smallest scale. In oxic or microaerobic sediments, particle aggregates such as faecal pellets of small animals or small detrital particles develop internal anaerobic and reducing conditions (Section 1.1). The particles are then surrounded by sulphide-oxidizing bacteria. These mm-large particles each then maintain a complete sulphur cycle. Such spatial arrangements allow for very high microbial activity due to the short diffusion pathways involved in the element cycling.

4.1.5 *Anaerobic deserts*

The most studied anaerobic habitats swarm with life; few microbial communities show higher densities (and often diversity) of prokaryotes and unicellular eukaryotes than sulphureta, the uppermost part of the anaerobic zone of shallow-water sediments or the rumen of a cow. This is not surprising since the requirement for maintaining anaerobic conditions within short distances of the aerobic world is a high import of organic material and a high biological activity (cf. Section 1.1). However, anaerobic deserts with a very low biomass and with a very low biological activity do occur. These belong to one of two categories: those with very low concentrations of mainly refractile organic substrates; and those which have become 'sour' so that biological degradation is blocked, although organic material is still abundant.

The first category is represented by the deeper layers of all aquatic sediments. Below the depth of bioturbation and of re-suspension of sediment there is no re-supply of organic substrates and only the refractile compounds remain. If microorganisms degrade these, they do so very slowly. Figure 4.16 shows the vertical distribution of sulphate and of the rate of sulphate

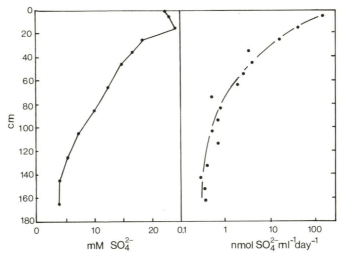

Fig. 4.16 The vertical distribution of sulphate and rates of sulphate reduction in the sediment down to a depth of 1.6 m in Limfjorden, Denmark (redrawn from Jørgensen 1977*b*).

reduction down to a depth of 1.6 m in an estuarine sediment (Jørgensen 1977b). More than two-thirds of the entire sulphate reduction takes place in the upper 10 cm of the sediment although it can still be detected at a depth of at least 1.6 m. This, first of all, demonstrates the extreme sensitivity of the radiotracer method for quantifying sulphate reduction. Its rate at this depth is about three orders of magnitude lower than at the surface and the turnover time of the sulphate pool is between 200 and 300 years. The sedimentation rate of this locality is not known with certainty, but the deepest part of the sediment core must represent the surface layers several centuries previously. (The sulphate-concentration gradient is not simple to interpret in this case due to compaction of the sediment and changes in salinity over historical time. Thus it cannot be used for accurate estimates of the rate of sulphate reduction in this sediment; see Jørgensen 1978b.)

This example is typical of aquatic sediments in general: below depths of 10–20 cm, biological activity and microbial biomass are very low. The habitat remains anaerobic because oxygen is consumed at the surface and because the diffusion pathway is very long. Eventually, abiological rather than biological processes become important in the diagenesis. Some organic material is never mineralized, but eventually becomes the kerogen of sedimentary rocks. Under high pressure and high temperatures such material may eventually be transformed into methane and mineral oil.

Other types of anaerobic deserts are represented by acid mires and bogs. Fermentation (in conjunction with cation-exchange properties of *Sphagnum* mosses and humic substances) results in acid conditions, the system eventually gets stuck and further mineralization stops. In addition, a substantial part (20–30%) of vascular plants consists of lignin, waxes and resins which are very resistant to degradation under anaerobic conditions (Westermann 1993). In bogs, the structure of plant materials and even animal tissue is preserved, a fact which is of great value to archaeologists. The material is referred to as peat, and over geological time it is transformed into lignite and eventually hard coal.

4.1.6 'Flow-through' anaerobic communities

By 'flow-through' communities we mean the sort of continuous (multispecies) cultures which occur in nature or which are designed by engineers. These systems have an inflow (with a dissolved or particulate substrate), a reaction container with microbial populations which are in a more-or-less steady state condition, and an outflow carrying waste products and excess organisms. Natural anaerobic flow-through systems are represented by the intestinal tract (or part of it) in animals which depend on the microbial fermentation of their food. Artificial systems include anaerobic sewage digestors. We will not deal with these systems in any detail here, but we will outline some general principles. Microbial intestinal communities were briefly discussed in Section 1.3, communities of fermenting bacteria (with examples from the rumen) were discussed in Sections 2.1 and 2.3 and the protozoan biota of these systems will be treated in Section 4.2. Otherwise the reader is referred to Hobson (1989) and Hungate (1966) regarding the rumen and to Schink (1988) and references therein regarding anaerobic sewage digestors.

The rumen is by far the best studied system. In the cow, the ingested feed passes together with saliva via the reticulum to the rumen. The rumen–reticulum system, which constitutes about 15% of the volume of the animal, is divided into four lobes. In the rumen, structural carbohydrates (cellulose, xylan, pectins, starch) are hydrolyzed and stoichiometrically fermented into acetate, propionate and butyrate (which together represent the carbon source of the animal), and methane (in addition to CO_2 and H_2O). The contents of the rumen then pass via the omasum (where the fatty acids and bicarbonate is absorbed) to the abomasum (true stomach) where acid digestion of the microorganisms takes place; the microorganisms constitute the only protein source for ruminants since all ingested proteins are de-aminated and fermented in the rumen.

The rumen contains a variety of bacteria plus protozoa and chytrids. Different types of fermentations take place and they include several intermediate products with a rapid turnover. One important intermediate is hydrogen which is removed by methanogens, especially *Methanobacterium ruminantium*. An important aspect of the anaerobic degradation of the structural carbohydrates, which represents the bulk of the feed, is the low energetic efficiency of fermentation. Thus, the total energy yield of the fermentations in the rumen is about 4 mol ATP per mole glucose (180 g) and this allows for the synthesis of about 40 g of microorganisms. About 10% is lost in the form of CH_4, but about 65% (by weight) ends up as fatty acids, which are available to the animal. A fermentative degradation of the food thus provides much organic carbon which can be used by the aerobic host animal. An interesting question is why propionate and butyrate are not to a larger extent transformed into acetate by proton-reducing acetogenic bacteria and why acetate is not transformed into methane; processes which would represent a loss to the ruminant. The reason may be the high turnover of the rumen content (15–50 h), which does not allow for the relatively slow growth rates associated with these processes.

Systems such as the rumen are often compared to a chemostat. In a chemostat, the microbes are assumed to be in an homogeneous suspension and to be subject to the same dilution rate as the bulk fluid which contains the dissolved substrate. For a chemostat, the equations describing the time dependent concentrations of cells and substrates have already been given

(equations [1] and [2] in Box 3.1). Recall that the equilibrium size of x, is given by:

$$\bar{x} = Y[S_r - DK/(\mu_m - D)], \tag{4.1}$$

and the equilibrium value for S is given by:

$$\bar{s} = DK/(\mu_m - D) \tag{4.2}$$

where S is the substrate concentration (in the reaction flask and in the effluent), Y the growth yield of x (increase in x per unit consumed S), S_r the substrate concentration in the reservoir (inflow), D the dilution rate, K the half-saturation constant, and μ_m is the maximum growth rate constant approached at very high S-concentrations. From these two equations it can be seen that if the rumen worked as a classical chemostat, it would be very inefficient, because only at very low dilution rates will a large part of the substrate be metabolized in the system. At higher dilution rates, the system will export a relatively large fraction of the substrate; S will be high in the reaction flask, supporting a rapidly growing, but small population of x.

The cow has solved this problem by using differential retention times for the liquid phase and the biota. In terms of the above equations, [4.1] and [4.2] (for the steady state situation) now become:

$$\bar{x} = YD_2/D_1[S_r - D_2K/(\mu_m - D_1)] \tag{4.3}$$

and
$$\bar{s} = D_1K/(\mu_m - D_1) \tag{4.4}$$

where D_1 and D_2 are the dilution rates for the biota and for the substrate flow, respectively. It can be seen that if $D_1 < D_2$ a larger equilibrium size of the microorganisms and a more efficient utilization of S will result. This effect is also illustrated in Fig. 4.17. In real chemostats this (usually undesirable) effect may result from 'wall-growth'. The cells tend to adhere and grow on the walls of the reaction flask so that they have a lesser tendency to be flushed out of the flask than if they were permanently suspended in the medium. As a result, the population sizes are larger, and the per capita growth rates and substrate concentrations lower, than predicted from [4.1] and [4.2]. The system is no longer a real chemostat and cannot be used for reliable estimates of growth parameters.

The actual mechanisms in the cow rumen have been studied by Czerkawski and Cheng (1989). The microbes tend to stick, more-or-less reversibly, to surfaces such as the wall of the rumen and those of particulate matter. Larger particles tend to have a longer residence time in the rumen (2–3 days, or about 2–3 times longer than the bulk fluid). This is accomplished through the compartmentalization of the rumen and muscular activity, so that larger particles (with attached microbes) are maintained longer in the rumen, where they partly disintegrate, but the rumen fluid is

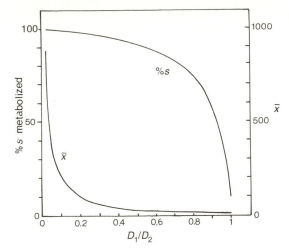

Fig. 4.17 A numerical example of the consequence of equations [4.3] and [4.4]. The parameter values are: $\mu_m = 1$, $K = 10$, $Y = 0.2$ and $D_2 = 0.9$. The dilution rate, D_1 [= (residence time of the microbes)$^{-1}$] is varied. The result shown is the population density ($\bar{x}$) and the percentage of the substrate which is metabolized. It is seen that in a normal chemostat ($D_1 = D_2$) only about 10% of the substrate is metabolized, but if the retention time of the microorganisms is prolonged, the efficency is increased.

regularly strained through the particle masses. An additional complexity is the fact that different types of bacteria tend to have somewhat different habitats within the rumen.

The rumen is thus considerably more complex than a simple chemostat. Its optimal function (maximation of fatty acid production) requires not only a control of the flow rate of the bulk contents, but also a differential residence time of the different components (solutes, particles and microbes). A complete model, which can explain the exact stoichiometry of the fermentation processes in terms of microbial processes as well as the hydraulic and mechanical properties of the rumen, has not yet been achieved.

Organic wastewater treatment is also based on microbial degradation; both aerobic and anaerobic systems are used. Anaerobic systems are generally considered to be inefficient for large scale wastewater treatment. Most cities therefore use trickling filters or the activated sludge process; in the latter system the 'sludge' (largely microbial biomass) is recycled to maintain a sufficiently high bacterial biomass (cf. the discussion on the rumen; p. 197). The advantage of the anaerobic sewage treatment is, however, that it produces less sludge, because the anaerobic mineralization yields less energy and hence a smaller production of microbes. An anaerobic digester is therefore often used in conjunction with the (aerobic) activated sludge process to mineralize excess sludge.

On a smaller scale, however, anaerobic sewage digestors are used and sometimes exploited for their production of methane. In contrast to the rumen, the goal is to achieve a complete mineralization (ideally with only CH_4 and CO_2 as end-products). The retention time is therefore much longer (2–3 weeks). As in the rumen, it is of paramount importance that the retention time of the microbes is higher than that of the liquid to ensure high population densities and a high efficiency of degradation of the substrates at relatively high turnover rates of the wastewater. Different designs have been used to achieve this including a settling tank for the effluent, recycling of the sludge, or filling of the reactor tank with gravel or other material to provide a large surface area for bacterial adhesion.

4.2 Anaerobic protozoan biota

We have already seen (Section 3.1) that anaerobic habitats are inhabited by representatives from a limited number of protozoan taxa and all anaerobic habitats show at least some similarities regarding the composition of their eukaryotic biota. Among the flagellated protozoa, diplomonads and retor- tamonads are found in virtually all anaerobic habitats which harbour eukaryotes, while representatives of the Parabasalia (trichomonads, hyper- mastigids) and the oxymonads occur almost exclusively as intestinal symbi- onts. Representatives of the rhizomastigids are known only as free-living species. Among the ciliates, representatives of the trichostomatids and the heterotrichids (plus the commensal entodiniomorphids and the odontosto- matids which are closely related to the trichostomatids and the heterotri- chids, respectively) dominate all anaerobic communities in which ciliates occur. Some other protozoan groups are also represented in certain anaero- bic habitats, but the above-mentioned groups are common to a wide range of different anaerobic communities. Certain anaerobic communities of symbiotic protozoa have enjoyed a long independent evolution which has resulted in amazingly specialized and species-rich assemblages. The most impressive examples are the intestinal biota of termites and wood-eating cockroaches (which must have evolved since the Carboniferous) and the protozoan communities of the intestinal tracts of herbivorous mammals (which must have evolved during the Tertiary). The anaerobic (non-symbi- otic) biota of limnic and marine habitats are characterized by their similarities in taxonomic composition and community structure, although there are also some differences.

With this background, it is natural to treat free-living and symbiotic anaerobic communities separately. There are also good historical reasons for doing so: the anaerobic nature and functional significance of the protozoan biota of herbivore intestinal tracts were understood relatively early. In contrast, the recognition of communities of free-living anaerobic eukaryotes is more recent (although most species were described long ago). While there are several recent articles and reviews treating various aspects of anaerobic symbiotic protozoa, literature on free-living anaerobic protozoa is scarcer; consequently these are treated in more detail here.

4.2.1 *Communities of free-living anaerobic protozoa*

The characteristic biotic communities which inhabit decaying organic material (such as polluted freshwater habitats and sewage treatment plants) were recognized early in this century (Kolkwitz and Marsson 1909; Lackey 1932; Lauterborn 1901; Liebmann 1936; Wetzel 1929). These habitats were referred to as 'sapropel' (from Greek, *sapros* = putrid and *pel* = sludge or mud) and the species were said to be 'sapropelic'. An extension of this terminology ('polysaprobic', 'mesosaprobic', etc.) still plays a role in the classification of organic pollution. The term 'sapropelic protozoa' has also been used as a synonym for anaerobic protozoa. This is unfortunate, however, since most of the organisms referred to in the literature as being sapropelic are not really anaerobes (but they are in many cases microaerophiles). Rather, they commonly thrive in habitats with high densities of bacteria, such as the aerobic film of trickling filters. Also, many genuinely anaerobic protozoa are not particularly associated with large amounts of accumulated organic matter. Some of the early studies on sapropelic communities recognized the association with low concentrations or the absence of oxygen, but until recently, the physiological and ecological implications of this was not studied in detail.

To marine biologists, the long-held conventional wisdom was that anaerobic habitats (such as sulphidic sediments) are 'azoic'. However, earlier studies on ciliate biota of sediments, salt marshes and sulphureta (e.g. Fauré-Fremiet 1951; Kirby 1934; Kahl 1928, 1931, 1930–35) did reveal the association between sulphidic environments and certain ciliate species, many of which are specialized for feeding on various kinds of sulphur bacteria. These studies also provided descriptions of most of the currently known anaerobic ciliates. Limnic 'sapropelic' ciliates were also given a comprehensive treatment in Kahl (1930–35). Descriptions of anaerobic species belonging to non-ciliate protozoan groups are found scattered in the older literature without any particular reference to their anaerobic nature (e.g. Grassé 1952). The existence of characteristic communities of free-living obligate anaerobic protozoa (as distinct from species which are tolerant to low p_{O_2} or transient exposure to anoxia or sulphide) was first recognized by Fenchel (1969) and Fenchel *et al.* (1977). There is great variation between anaerobic habitats with respect to the quantitative and qualitative richness of protozoan biota, and some habitats do not harbour eukaryotes at all. Protozoa may be absent for several reasons. Isolated anaerobic microniches, such as anaerobic soil particles or detrital particles suspended in oxic water may be too isolated, small or ephemeral to sustain protozoan populations. In sediments, the presence of protozoa is limited by the size of the interstitia. Thus, Fenchel (1969) found that ciliates are virtually absent in sands with a median grain size of less than about 125 µm. Smaller protozoa (flagellates, amoebae) do occur in more fine-grained sediments, but even these organisms will be absent in compacted silty and clay-containing sediments. The production of bacteria is so low in many anaerobic habitats ('anaerobic deserts', see Section 4.1) that they cannot sustain protozoan biota. In sandy sediments, numbers of individuals and species are highest immediately below the oxic zone where the bacterial activity is highest, and protozoan numbers decline rapidly with depth. In the anaerobic water column of stratified lakes or fjords the numbers increase towards the bottom, where sedimenting organic matter accumulates. An exception is the protozoan biota associated with the mass occurrence of phototrophic bacteria in the uppermost part of the anaerobic water column. The richest assemblages of anaerobic protozoa are found in sulphureta or other accumulations of decaying organic material, in sandy shallow-water sediments, and in the anaerobic zone of very productive freshwater systems.

Certain (or perhaps most) major groups of protozoa are not represented in anaerobic habitats. These include the radiolarians (which are characteristic of oceanic plankton), the heliozoans, most groups of naked and all testate amoebae, some groups of flagellates and 'fungi-like protozoa' (water moulds, myxomycetes, etc.). Living foraminifera have been observed in anaerobic (or at least 'dysaerobic') sediments; however, while some species are tolerant to longer or shorter exposure to anoxia, real anaerobic species do probably not exist (Bernhard 1989; Bernhard and Reimers 1991). Anaerobic chytrids are known from the rumen, and free-living species could be expected to exist, but they have not yet been recorded. It is likely that some groups, especially those consisting of very small species, may still be found to have anaerobic representatives. Thus, Patterson and Fenchel (unpublished observations) found that some representatives of flagellate groups which are otherwise not associated with anoxic habitats (stramenopiles, heteroloboseans, bodonids and choanoflagellates) do multiply in anaerobic crude cultures inoculated with marine sediments.

The most conspicuous and quantitatively most important groups of protozoa in anaerobic limnic and marine habitats include the flagellated diplomonads and retortamonads, the amoeboid rhizomastigids (Fig. 4.18) and certain groups among the ciliates (Fig. 4.19 and 4.20); in addition, a few organisms which have so far not been studied in detail, or organisms

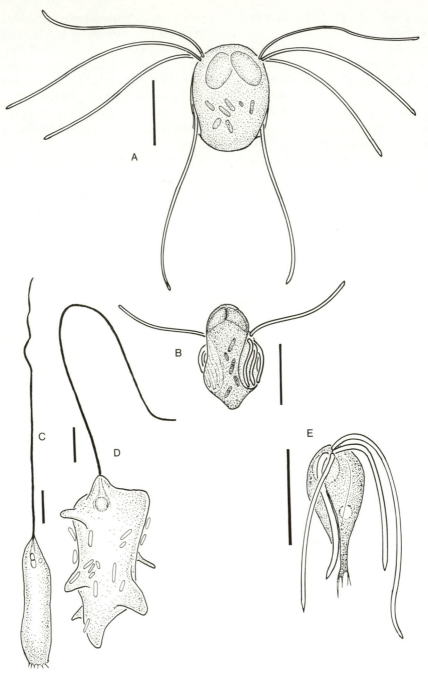

Fig. 4.18 A–B: diplomonad flagellates (A, *Hexamita inflata*; B, *Trepomonas agilis*). C–D, rhizomastigids (C, *Mastigamoeba* sp.; D, *Mastigella* sp.); E, the retortamonad *Chilomastix* sp. All species have been isolated from anaerobic marine sediments. Scale bars: 5 μm.

with a taxonomically uncertain position have been recorded (e.g. the flagellate *Psalteriomonas lanterna*; see Broers *et al.* 1990).

Among the three first-mentioned groups it seems (as least on the basis of morphological criteria) that many species are common to fresh- and

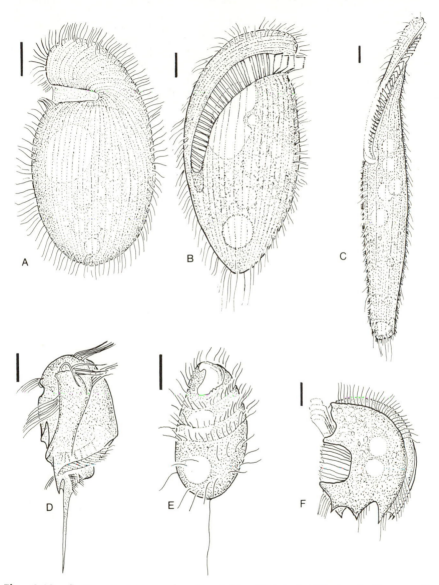

Fig. 4.19 Some common marine anaerobic ciliates. A, *Plagiopyla frontata*; B, *Metopus contortus*; C, *Parablepharisma* sp.; D, *Caenomorpha levanderi*; E, *Trimyema* sp.; F, *Saprodinium halophila*. Scale bars: 10 μm.

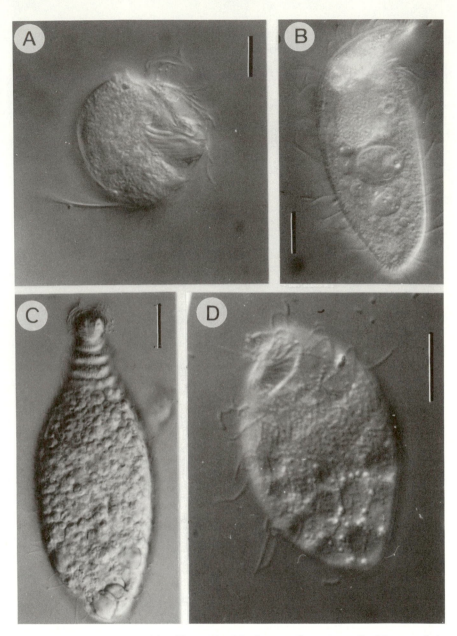

Fig. 4.20 Some anaerobic ciliates; A, *Myelostoma bipartitum*; B, *Metopus halo-phila*; C, *Lacrymaria elegans*; D, *Trimyema* sp. A–B are marine, C–D are limnic species. Scale bars A,B,D: 10 μm, C: 20 μm.

seawater. However, among the rhizomastigids, at least the large *Pelomyxa palustris* is confined to freshwater, and it may well be that small flagellates (such as e.g. *Hexamita inflata* or *Trepomonas agilis*) include freshwater and marine strains which are indistinguishable from each other. The species compositions of marine and freshwater ciliate biota are different, although closely related congeners are found in both. Two anaerobic representatives of the genus *Lacrymaria* (*L. elegans* and *L. sapropelica*, Fig. 4.20), however, seem to occur in sea- as well as in freshwater (although a closer examination may reveal the existence of different marine and limnic forms). A strain of *Metopus palaeformis*, originally isolated from a landfill site (Finlay and Fenchel 1991*a*), will grow in cultures based on freshwater as well as on 50% seawater (unpublished observation).

Recent studies (Esteban *et al.* 1993*a,b*; and unpublished observations) have revealed a number of previously undescribed anaerobic ciliates in freshwater suggesting that some species still remain to be found. With the exception of some genera containing very small forms, however, it is not a frequent event to find undescribed species and it is likely that most have been named. With this background it is possible to make some generalizations concerning differences between limnic and marine biota. On the basis of the literature and with the assumption that certain taxa consist exclusively of anaerobes, there seems to be a higher number of anaerobic freshwater than marine species. Among the metopids (*Metopus* and *Brachonella*) about 60 species have been described and among these only 6 are marine (plus 2 undescribed species). Similarly, about 15 caenomorphids (*Caenomorpha*, *Ludio*, *Cirranter*) have been described, of which 2 are marine. The order Odontostomatida (all of which are anaerobes) comprise nearly 30 species, but only 2 are marine. On the other hand, the genus *Parablepharisma* comprises only (about 5) marine species. Among the trichostomatids, the genus *Plagiopyla* includes 3 marine and 3 limnic forms, while the genus *Sonderia* is exclusively marine, with 8 named species. Seven (mostly marine) species of the genus *Trimyema* have been named (Augustin *et al.* 1987), but the genus seems to include several undescribed species in both sea- and freshwater (Finlay *et al.* 1993; unpublished observations). Other groups of ciliates which have (usually few) representatives in anaerobic, limnic or marine habitats include haptorids, prostomatids, scuticociliates and oligotrichs. They are to some extent represented by different genera in marine and limnic habitats, respectively, but no generalizations regarding the relative species-richness can be made. In our own studies we have found about 60 free-living species of obligate anaerobic ciliates (Table 3.4), with roughly equal numbers of limnic and marine forms.

Large aerobic protozoa which occur in the water column (especially in the sea, but to some extent also in freshwater) are specialized plankton forms which show a number of characteristic adaptations. These include the ciliated tintinnids, many dinoflagellates and the radiolarians. In contrast, the protozoan biota of the anaerobic water column seem to be a subset of benthic forms, although the protozoa of large anoxic water bodies have not yet been studied in detail. (The absence of specialized planktonic species also applies to the microaerophilic protozoan biota which occur in the oxycline above the anoxic water column). Populations of scuticociliates, possibly belonging to the genus *Cristigera*, the anaerobic representatives of which carry ectosymbiotic bacteria, have been recorded beneath the oxycline of the Black Sea (Zubkov *et al.* 1992). Somewhat more detailed studies in two Danish fjords with an anoxic water column revealed, in addition to small heterotrophic flagellates, populations of scuticociliates and of *Metopus contortus*, *Plagiopyla frontata*, *Caenomorpha levanderi* and a few other species typical for anaerobic sediments (Fenchel *et al.* 1990; see also Fig. 4.21). Studies of the protozoan biota of the anaerobic hypolimnion of some Spanish meromictic lakes (Esteban *et al.* 1993*b*; Finlay *et al.* 1991; unpublished observations; see also Fig. 4.22) demonstrated the presence of relatively species-rich ciliate assemblages, but again, the recorded species can also be found in detritus and in sediments. These studies also revealed different food niches (based mainly on food particle size), results which can to some extent be generalized to other anaerobic ciliate biota. The odontostomatids and the scuticociliates depend on the smallest bacteria for food, while caenomorphids (*Caenomorpha*, *Ludio*) feed mainly on the larger, photosynthetic bacteria (Guhl and Finlay 1993). The *Holophrya* sp. seems to feed on sinking *Cryptomonas* cells (which derive from dense populations

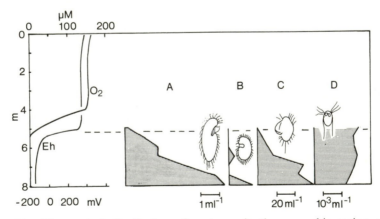

Fig. 4.21 The vertical distribution of protozoa in the anaerobic region of the stratified water column in Hjarbæk Fjord, Denmark, July 1986. A, *Metopus contortus*; B, *Plagiopyla frontata*; C, scuticociliates; D, heterotrophic flagellates. The flagellates were not identified and therefore some aerobic forms are included around the chemocline (data from Fenchel *et al.* 1990).

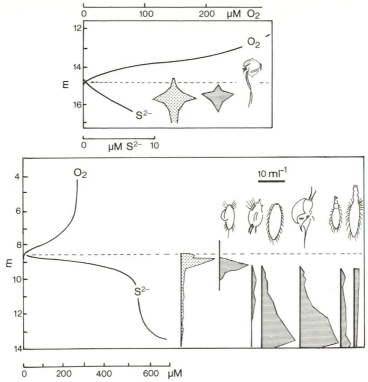

Fig. 4.22 Vertical distribution of anaerobic ciliates in the water column of two Spanish meromictic lakes. *Above*: Laguna de la Cruz, September 1987. Distribution of bacteriochlorophyll *a* (dotted; maximum abundance: *ca* 125 μg/ℓ and the ciliate *Ludio parvulus* (maximum concentration: *ca* 300 cells/ml) which feeds on the purple bacteria (B. J. Finlay, unpublished data). *Below*: Arcas II, August 1990. The vertical distribution of purple sulphur bacteria (dotted; maximum concentration 2×10^6 cells/ml) and from left: scuticociliates (*Cyclidium dilectissimum*, *Isocyclidium globosum*, and *Cristigera* sp.), odontostomatids (*Epalxella striata* and *Pelodinium reniforme*), *Holophrya* sp., *Caenomorpha medusula*, *Lacrymaria sapropelica* and *L. elegans*) (based on Finlay *et al.* 1991 and Esteban *et al.* 1993*b*)

in the metalimnion). The *Lacrymaria* spp. are probably mainly predators of heterotrophic flagellates.

Similar biota occur in other types of anaerobic limnic habitats such as flocculent anaerobic sediments and accumulations of decaying leaves (Finlay *et al.* 1988). Here, other types (especially species belonging to the genera *Metopus*, *Plagiopyla* and *Trimyema*) are also important. The giant amoeba *Pelomyxa palustris* is one of the most peculiar inhabitants of the anaerobic or microaerobic surface sediments of ponds and lakes. Like other rhizomastigids it is devoid of mitochondria, but it contains three types of endosymbiotic bacteria, two of which may be methanogens (Van Bruggen *et al.* 1983; Whatley and Chapman-Andresen 1990). Since it has not been grown in pure culture its energy metabolism is not well understood. It phagocytizes almost all types of organic particles and it seems to undergo a complex polymorphic life cycle during which it is partly oxygen-intolerant and partly microaerophilic (Whatley and Chapman-Andresen 1990).

The richest marine anaerobic protozoan biota are found in and on productive sandy shallow-water sediments. Sulphureta and microbial mats (see 4.1) are especially rich and may harbour several hundred ciliates per ml in addition to flagellates and amoebae, while the interstitia in the deeper parts of sandy sediments harbour lower numbers of protozoa (Fig. 4.23). Among the ciliates, several species of *Metopus*, *Plagiopyla frontata*, *Caenomorpha levanderi*, *Cristigera* spp. and two odontostomatid species are important bacterivorous forms. The *Parablepharisma* spp. are also common, among which the largest, *P. pellitum* is specialized for feeding on larger bacteria such as purple sulphur bacteria. Species of *Sonderia* feed on large sulphur bacteria including filamentous forms (Fig. 5.1C). As in freshwater, ciliates belonging to the genus *Lacrymaria* and possibly a few other haptorid and prostomatid ciliates represent the only predators of other protozoa. Bacterivorous diplomonad flagellates and rhizomastigids (*Mastigamoeba*, *Mastigella*) also occur, but flagellates and amoeboid organisms have not been studied in much detail. The ciliate *Trimyema* and the diplomonad

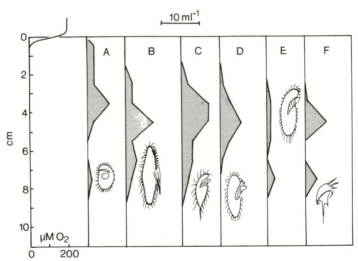

Fig. 4.23 The vertical distribution of larger anaerobic ciliates in a sandy sediment, Nivå Bay, May 1992. A, *Sonderia* spp.; B, *Parablepharisma pellitum*; C, *Metopus vestitus* + *M. halophila*; D, *Metopus* sp.; E, *Metopus contortus*; F, *Caenomorpha levanderi* (based on Fenchel 1993).

Hexamita seem especially to occur during the initial degradation of large amounts of accumulated organic material (Fenchel 1968, 1993; Dando *et al.* 1993).

4.2.2 *Communities of intestinal protozoa*

Most animals harbour several species of intestinal protozoan symbionts; these are mainly harmless commensals, but some are more or less pathogenic and some are useful to the host. Not all of these protozoa are anaerobic species, but in larger animals and especially in herbivores, part of the intestine does constitute an anaerobic or almost anaerobic habitat due to high bacterial activity (see Section 2.1). Protozoa including representatives of the trichomonad, diplomonad and retortomonad flagellates, and often the trichostomatid (especially *Balantidium*) and heterotrichid ciliates, occur in a wide variety of vertebrate and invertebrate guts. Man harbours species of the amoeboid *Entamoeba*, the diplomonad *Giardia*, the parabasalian *Trichomonas*, the retortomonad *Chilomastix* and the ciliate *Balantidium coli* (one *Entamoeba* species and *Giardia* may be pathogenic; the ciliate, which also occurs in pigs, is quite rare in man). The anaerobic nature of intestinal inhabitants has not been established in all cases (cf. discussion on intestinal nematodes, pp. 165–6). Even forms which are known to lack oxidative phosphorylation are often quite oxygen tolerant (e.g. the *Trichomonas* and *Giardia* spp; see Section 3.3). This is probably an adaptation for surviving the transfer from one host to another; also, microaerobic conditions occur in all animal guts. Some *Trichomonas* species occur in aerobic sites within the host (e.g. in the human oral cavity and vaginal tract). As far as energy metabolism and biochemistry is concerned, the species which occur in man are the most closely studied anaerobic protists (Section 3.3).

The number of protozoan species which have been recorded from the guts of different animals is impressive as can be seen from texts on protozoology (e.g. Grassé 1952 or Kudo 1966). The physiological properties and the functional significance of these organisms, however, have rarely been studied. The relatively species-rich ciliate biota of regular sea-urchins (see Corliss 1979) are of some interest in that they include representatives of some anaerobic genera (*Metopus* and *Plagiopyla*) which otherwise include free-living species. This shows how the intestinal biota were originally introduced by free-living forms which were accidently ingested together with decaying material or anaerobic sediment. Berger and Lynn (1992) have recently shown that the *Plagiopyla* (and the related *Lechriopyla*) from sea-urchins have hydrogenosomes and symbiotic methanogens similar to those of their free-living relatives.

The most interesting and most closely studied examples are the intestinal biotas of herbivorous mammals and of wood-eating insects. In both cases an astonishing diversity has evolved and the number of species probably greatly exceeds that of non-symbiotic, anaerobic protozoa. The symbiotic relationship with protozoa is vital to most groups of termites and cockroaches; herbivorous mammals are also entirely dependent on microbial fermentation, but the function of the protozoan component of the microbial biota is still incompletely understood: ruminants invariably remain healthy after experimental removal of the protozoa.

In wood-eating insects (cockroaches and most termite families) cellulose degradation depends on symbiotic anaerobic flagellates. These belong to the Parabasalia (the trichomonads and the hypermastigids — the latter occurring only in these insects) and the Oxymonadida which are probably related to the parabasalians and which also occur exclusively in the hindgut of wood-eating insects. The great diversity of these forms and their biology has been documented especially by Cleveland, Grassé, Hungate and Kirby; for references see Grassé (1952), Hungate (1955) and Kudo (1966). More recent reviews on the biology of these organisms are given by Breznak (1984) and Honigberg (1970); see also Fig. 4.24.

All three groups of flagellates have evolved large and complex forms which are often multinucleate and multiflagellate. They also have complex life-cycles which may involve attached and free-swimming stages, sexual phenomena and encystation. The induction of the different life cycle stages is partly controlled by host hormones which induce moulting or reproduction of the insects. Thus, termite flagellates encyst prior to moulting; the cysts remain in the excuviae which are then eaten by the termites to secure re-infection (the hindgut wall is a part of the ectoderm and it is discarded during moulting). In the living termite, the flagellates are found densely packed in the hindgut and they may account for 15–30% of the total insect volume (Hungate 1955).

The flagellates are often associated with both ecto- and endosymbiotic bacteria; in some cases the ectosymbionts are spirochaetes which contribute to the motility of the host. The functional significance of most of the associated bacteria is unknown; some may play a role in cellulose degradation and in some cases they are methanogens, which utilize the hydrogen produced by the fermentative host metabolism (Lee *et al.* 1987).

The flagellates contain ingested pieces of wood and they are capable of degrading cellulose, fermenting it to volatile fatty acids which are absorbed by the host. The host is entirely dependent on this for its nutrition and de-faunated termites starve to death unless they are re-infected.

Herbivorous mammals with pre- or postgastric fermentation also harbour species-rich and complex protozoan communities. These include flagellates (retortamonads, diplomonads, trichomonads) as well as chytrids which are known to play a significant role in carbohydrate fermentation. The most conspicuous component, however, is the community of ciliates. Pioneering

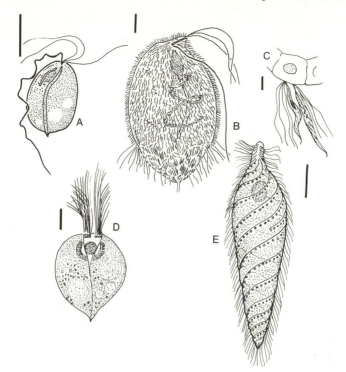

Fig. 4.24 Parabasalian flagellates. A–B: trichomonads (A, *Trichomonas caviae* from the guinea-pig; B, *Caduceia theobromae* from termites; the flagellate bears two species of ectosymbiotic and one species of endosymbiont bacteria); C, the oxymonad *Streblomastix strix* attached to the gut wall of a termite. D–E: hypermastigids (D, *Lophomonas blattarum* from a cockroach; E, *Spirotrichonympha flagellata* from termites). Scale bars: 10 µm. (redrawn from Grassé 1952).

studies of the biology of the rumen (which has been studied in much more detail than has the caecum of mammals with postgastric fermentation) were carried out by Hungate (1955, 1966, 1975); recent reviews (with special emphasis on protozoan communities) are those of Coleman (1980), Hobson (1989), Williams (1986) and Williams and Coleman (1991). The diversity of ciliates from the gut of herbivorous mammals is also documented in Corliss (1979) and Kudo (1966).

Among the ciliate orders present are the prostomatids (Buetschlidae with several genera) and suctorians (*Cyathodinium, Arcosoma*) which are known mainly from the caecum of perissodactyls and rodents. More important, however, are representatives of the trichostomatids and of the entodiniomorphids (the latter probably being derived from the former). Members of the trichostomatid families Pycnotrichidae and Paraisotrichidae occur mainly in perissodactyls, in hyrax, elephants and rodents, but also in camels and ruminants. The isotrichidae (*Dasytricha, Isotricha*, see Fig. 4.25) occur in ruminants where they constitute a substantial part of the ciliate community. These ciliates play a considerable role in fermenting carbohydrates including starch, but not cellulose. The entodiniomorphids constitute the most diverse group of ciliates in these habitats and they are found in the rumen as well as the caecum of species with post-gastric fermentation; one genus (*Troglodytella*) occurs in the chimpanzee and the gorilla. Altogether, 42 genera are recognized; common genera in the rumen of cattle and sheep are species of *Entodinium, Polyplastron* (see Fig. 4.25) and *Ophryoscolex*. Their biology is diverse and includes forms which can degrade and ferment

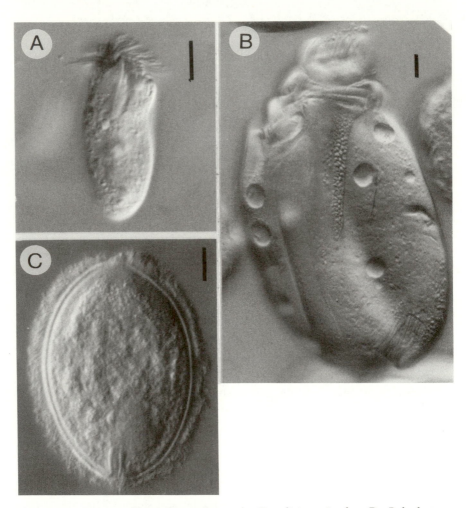

Fig. 4.25 Rumen ciliates from sheep. A, *Entodinium simplex*; B, *Polyplastron multivesiculatum*; C, *Dasytricha ruminantium*. All scale bars 10 μm.

cellulose and other structural carbohydrates (and which are always found to contain ingested fragments of plant tissue), bacterivores and forms which ingest other protists.

The density of ciliates in the rumen is high (> 10^5 cells ml^{-1}) and they constitute a substantial part of the microbial protein on which the host depends. There has been a protracted debate on the question of the necessity of rumen protozoa and whether or not they contribute to efficiency of fodder utilization in ruminants. Since young ruminants are initially inoculated with their microbial biota by ingesting fresh sputum deposited by older ruminants on the grass or leaves, it is easy to produce ruminants which are devoid of protozoa or even the entire microbial biota. Extensive experimentation (reviewed in Williams and Coleman 1991) has shown that while the bacterial biota are absolutely necessary if the ruminant is to remain a herbivore, the absence of protozoa has little if any detrimental effect. The reason for this is that the bacterial density of the rumen increases in the absence of protozoan grazing and so the amount of microbial protein available to the mammal remains unchanged.

4.3 The role of phagotrophs in anaerobic communities

The role of free-living eukaryotic phagotrophs in anaerobic habitats was neglected until quite recently. This can be explained by the fact that they are often not particularly numerous, and the biomass they represent is small relative to that of their bacterial prey (as compared to aerobic habitats). This is a consequence of the low bioenergetic efficiency of anaerobic dissimilatory metabolism.

Rapidly growing animal tissue and aerobic protists have a net growth efficiency of around 60% (Calow 1977; Fenchel and Finlay 1983). This can be rationalized by the following considerations assuming that the generation of 1 mol ATP is sufficient for the synthesis of 4 g cell carbon (Bauchop and Elsden 1960). As discussed in 2.1 this is not a universal constant and Y_{ATP} can vary according to growth rate and substrate, but we will accept a value of 4 g C per mol ATP as a likely figure. If aerobic eukaryotic cells dissimilate 1 mol glucose (72 g C) they can generate 32 mol ATP which allow for the synthesis of 4×32 g organic C. They must therefore ingest $4 \times 32 + 72$ g organic C and the consequent net growth efficiency becomes $4 \times 32/(4 \times 32 + 72) = 0.64$. Phagotrophic cells are rarely capable of digesting 100% of what they ingest; the digestion efficiency probably varies widely according to the nature of the food, but a reasonable average would be 70%. In that case, the organism would dissimilate 30%, assimilate 40% and lose 30% of what it eats, and the gross growth efficiency (yield) would be about 40%.

A fermenting protozoon will generate only 2–4 mol ATP per mol dissimilated glucose. Accepting 3 ATP/glucose as a likely average (2.1 and 3.3) we find a net growth efficiency of $4 \times 3/(4 \times 3 + 72) = 0.14$. There is no reason to assume any difference in digestion efficiencies of aerobic and anaerobic protozoa and so, using the same argument as above, we arrive at a yield of about 10% for an anaerobic protozoan, or about one-quarter of that of aerobes. There is also no reason to assume that the efficiency of food particle collection of anaerobic ciliates should differ from that of otherwise similar aerobic species, so we would expect the minimum generation time of anaerobes to be on average four times longer than that of similarly sized aerobes. Figures 4.26 and 4.27 present evidence which supports these generalizations. These quantitative differences between aerobic and anaerobic phagotrophic organisms result in differences between aerobic and anaerobic community structures.

One of these differences is that anaerobic food chains must be short because such a relatively large fraction of the ingested food is dissimilated at each level of the food chain. In fact, a second trophic level (organisms feeding on bacterivorous protozoa) is rare in anaerobic communities and

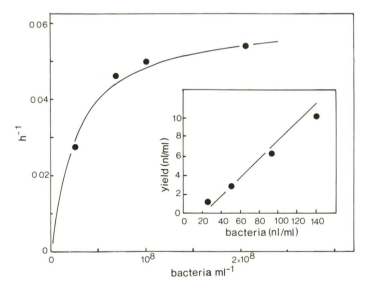

Fig. 4.26 The exponential growth rate constants of a marine isolate of the ciliate *Trimyema* sp. during the initial period of growth in batch cultures (25 °C) as a function of the initial concentration of food bacteria (*Pseudomonas* sp.); the data are fitted to Monod kinetics. *Insert*: eventual yield of ciliates as a function of the initial amount of bacteria (in terms of volume). The maximum growth rate (≈ 0.055 h^{-1} corresponding to a generation time of 12.5 h) and the growth yield (≈ 0.1) are both about one-quarter of that expected for a similarly sized aerobic ciliate (T. Fenchel, unpublished data).

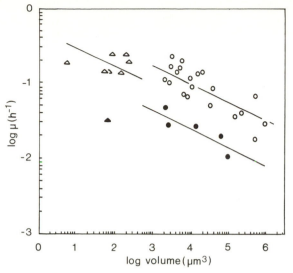

Fig. 4.27 Approximate maximum growth rate constants (20 °C) for some aerobic (open symbols) and anaerobic (closed symbols) protozoa as a function of cell volume. (triangles: flagellates, circles: ciliates). The maximum growth rate constant scales with (cell volume)$^{-0.25}$; it is seen that the maximum growth rate constant of the anaerobes is about one-quarter of that of aerobes of similar size (data from Fenchel and Finlay 1990*a*).

even higher trophic levels do not exist. This is discussed further in Section 5.3. A related consequence is that the expected ratio between predator and prey biomass in anaerobic communities should be only one fourth of that of aerobic communities. Fenchel and Finlay (1990*a*) compared the ratios between the bio-volume of bacteria and bacterivorous protozoa from the aerobic and the anaerobic parts of the water column (a stratified marine fjord and some stratified ponds and lakes). In all cases the predator/prey biomass ratio was lower in the anaerobic part (on the average by a factor of 0.24, ranging from 0.14 to 0.36).

All this does not, however, suggest that the effect of phagotrophs on anaerobic communities of bacteria is negligible, only that relative to aerobic microbial communities, a much larger part of the food of the phagotrophs is dissimilated (and the metabolites are then used by sulphate-reducing or methanogenic bacteria). Bacterivorous protozoa do affect the population sizes of bacteria in anaerobic communities and, due to the food particle size selectivity of different types of protozoa, the size range of the bacterial populations.

Figure 4.28 shows the microbial succession based on organic material incubated anaerobically in seawater and inoculated with a small amount of marine sediment. The course of events resembles that of similar aerobic

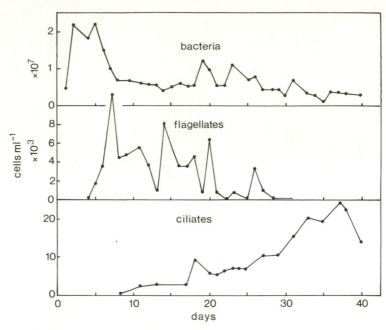

Fig. 4.28 The succession of bacteria, heterotrophic flagellates (mainly *Trepomonas* and *Hexamita*) and ciliates (initially *Cristigera* sp. and later *Plagiopyla frontata* dominated) in a flask with $10\,\ell$ anaerobic seawater (under a N_2-atmosphere) with boiled lettuce leaves (0.07 g C/ℓ) and inoculated with 5 ml marine sediment. After 40 days the organic matter was almost totally mineralized, as measured by the disappearance of SO_4^{2-} (T. Fenchel, unpublished data).

successions: initially, bacteria proliferate rapidly, but their numbers subsequently decrease due to grazing by increasing numbers of small protozoa (heterotrophic flagellates). These are eventually displaced by larger protozoa (ciliates); partly because the larger bacteria tend to escape predation by flagellates, because the larger ciliates ingest large bacteria more efficiently, and partly because the ciliates consume the smaller flagellates. A similar experiment under aerobic conditions would produce a higher biomass of secondary consumers, a higher diversity of species including protozoa which are specialized in preying on other protozoa, and the eventual occurrence of small metazoa, such as harpacticoids, nematodes and turbellarians.

5

Interactions with the oxic world

Anoxic environments do not exist in isolation from their oxic surroundings. As emphasized in Chapter 1, reducing power for the maintenance of anoxic environments, and the chemical energy needed by their inhabitants is, with a few exotic exceptions, provided by the oxic world in the form of organic material produced by oxygenic phototrophs. This situation has prevailed throughout most of the history of the Earth. Moreover, oxygen is necessary for the completion of biogeochemical element cycles; thus, nitrate and sulphate are regenerated from ammonia and sulphide through oxidation with O_2 (although some sulphide oxidation is carried out by phototrophs under anaerobic conditions). In turn, NO_3^- and SO_4^{2-} are needed under anaerobic conditions as electron acceptors in respiration. Anaerobic life is therefore most rich in the vicinity of aerobic habitats; anoxic environments which are relatively isolated from the aerobic world in terms of transport rates of organic substrates and chemical species, tend to be 'deserts'. Indeed, some of the most active anaerobic habitats in surface sediments occur as islands in a microaerobic matrix or they are only temporarily anaerobic such as the surface layers of microbial mats. In a world where molecular diffusion limits biological activity, physical proximity between aerobes and anaerobes in syntrophic interactions (e.g. sulphate reducers and sulphide oxidizers) is of great importance in an analogous fashion to syntrophy involving only anaerobic species (see Sections 2.3 and 4.1).

It is therefore natural to include a brief discussion on the boundaries between aerobic and anaerobic habitats and this will be the topic of the first section of this chapter. The following section will briefly review those biogeochemical processes which are peculiar to anaerobic habitats. Although we take only a qualitative approach to this topic, it will be shown that several properties of the contemporary biosphere depend on the existence of anaerobic biota.

In the final section we discuss the role of oxygen for the evolution of life. The emergence of food chains beyond two (or at most three) trophic levels required aerobic respiration for bioenergetic reasons. The evolution of large multicellular animals, however, required not only the microaerobic conditions which are sufficient for aerobic microbes (and for mitochondria), but also a high p_{O_2} approaching that of the present atmosphere. This explains the appearance of metazoa in the late Proterozoic and the subsequent 'Cambrian explosion' of invertebrate taxa.

5.1 Life at the oxic–anoxic boundary

5.1.1 *The chemolithotrophs*

The reduced end-products of anaerobic mineralization represent potential chemical energy in the presence of oxygen (see Fig. 1.2). Organisms which oxidize inorganic, reduced compounds (including C_1-compounds) are referred to as chemolithotrophs or chemoautotrophs; in most cases they utilize the energy they gain for CO_2-reduction via the Calvin–Benson cycle (for a detailed treatment of the biology and biochemistry of chemoautotrophs and for further references, the reader is referred to Schlegel and Bowien 1989).

Some of the reduced metabolites of anaerobes (notably H_2 and in part CH_4) are oxidized under anaerobic conditions by sulphate reducers, denitrifiers and methanogens. Although aerobic H_2-oxidizers exist, they are believed to play a small role in nature. Cyanobacteria produce some H_2 as a by-product of N_2-fixation and this may provide a niche for aerobic H_2-oxidizers in the surface of sediments (Schlegel 1989).

Microbes also catalyse the oxidation of reduced manganese and iron; Fe^{2+}-oxidation often leads to conspicuous rusty deposits on sediments of lakes and streams and also to visible layers of deposited MnO_2 in some marine sediments. In most cases these processes are believed not to be coupled to energy conservation. Ferrous iron oxidizes spontaneously and very rapidly in the presence of oxygen under neutral conditions and the resulting ferric hydroxides tend to deposit on the surfaces or in the mucous sheaths of certain bacteria such as the 'iron bacteria' *Gallionella* and *Leptothrix*. The only well-established example of ferrous iron being utilized as a substrate for respiration is that of *Thiobacillus ferroxidans*. This organism is found under acid conditions (where Fe^{2+} is more stable) and it plays an important role in pyrite oxidation in mine water or in drainage channels. Iron oxidation is coupled to electron transport phosphorylation (Kuenen and Bos 1989). In one case, there is evidence of a bacterium which can grow on the energy provided by Mn-oxidation (Kepkay and Nealson 1987).

The oxidation of ammonia, which results from de-amination of amino acids during the fermentation of amino acids, is an exclusively aerobic process. It takes place in the oxic–anoxic boundary layer. It is really a two-step process, in so far as two different groups of bacteria oxidize NH_4^+ to NO_2^- (*Nitrosomonas, Nitrosococcus*) and NO_2^- to NO_3^- (*Nitrobacter, Nitrococcus*; cf. Figs. 4.3 and 4.4).

That part of the CH_4 produced in sediments which does not escape to the atmosphere by ebullition is oxidized in the upper aerobic zone. If there is a relatively thick sulphate-containing layer above the methanogenic zone, however, anaerobic methane oxidation becomes important.

Sulphide may be oxidized anaerobically, not only by phototrophs, but also through denitrification by the chemolithotrophic *Thiobacillus denitrificans*; this is probably not a quantitatively important process in nature. In the dark, sulphide, thiosulphate and elemental sulphur are oxidized primarily by chemoautotrophic sulphur bacteria with oxygen as the electron acceptor. Most of these bacteria store sulphur (in the form of polysulphides) so long as sulphide is available. They therefore appear white in reflected light and they are often referred to as 'white sulphur bacteria'.

Sulphide oxidation is the quantitatively most important process in most systems. Sulphide oxidizes spontaneously in the presence of oxygen. The rates at which this proceeds in natural waters vary greatly according to sulphide and oxygen concentrations, pH and the presence of catalysts (such as iron); and published estimates of the half-life of sulphide in oxygenated seawater varies from minutes to more than 24 hours (Millero 1986). Even in the absence of biologically catalysed sulphide oxidation, the zone of overlap between sulphide and oxygen must be narrow in systems where solute transport is due solely to molecular diffusion.

The white sulphur bacteria are 'gradient organisms'; most depend on chemosensory motile behaviour in order to accumulate where there is an optimal (chemically unstable) mixture of O_2 and HS^-. These organisms may be motile filaments (such as *Beggiatoa* spp., Fig. 5.1) which occur mainly in sediments, or they may be attached filaments which occur as 'aufwuchs' on solid surfaces (*Thiothrix* spp.). *Thiovulum* is a very rapidly swimming and unusually large spherical bacterium. The cells form characteristic white 'veils' which enclose high concentrations of sulphide (Fig. 5.2). *Thiobacillus* spp. are smaller and do not store elemental sulphur. As in eukaryotic microaerophiles, the chemosensory behaviour seems to be controlled by p_{O_2}, rather than sulphide or other correlates of the chemocline. When conditions (mainly a sufficiently high upward flux of sulphide) permit high densities of *Beggiatoa*, the filaments form a 200–500 μm thick bacterial plate. Measurements with oxygen electrodes show that the maximum p_{O_2} tolerated by *Beggiatoa* (at the top of the bacterial film) is about 5% atm.sat. while the lower part is anaerobic. Since the bacterial filaments are continuously moving, they are probably exposed to anoxia for only limited periods of time; they are believed to subsist on heterotrophy (fermentation or sulphur

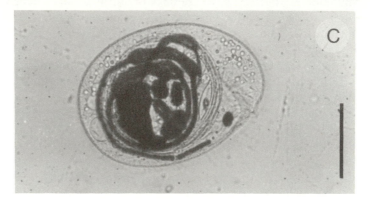

Fig. 5.1A, a macroscopic view of the sediment surface (Nivå Bay, August 1993) covered by a *Beggiatoa* mat. The decaying seagrass leaves have a width of about 0.5 cm. B, core sample of a *Beggiatoa* mat; the individual filaments can be seen. Scale bar: 1 mm. C, the ciliate *Sonderia schizostoma* with ingested filamentous bacteria, most of which are *Beggiatoa*; scale bar: 50 μm (C is from Fenchel 1968).

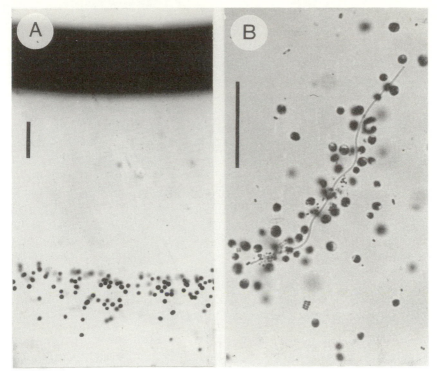

Fig. 5.2 A, The white sulphur bacterium *Thiovulum majus* forming a band in an oxygen gradient about 0.5 mm from the meniscus in a capillary tube. B, *Thiovulum* cells accumulating around a photosynthesizing cyanobacterial filament in an other wise anoxic microscopical preparation. Scale bars: 100 µm.

respiration) in the absence of oxygen (Nelson 1989; Nelson *et al.* 1986*a,b*). Such bacterial films are common on the surface of sulphidic sediments. They form conspicuous white patches or a continuous white sheet which may cover great areas of the sea bottom (Fig. 5.1). They occur frequently on shallow, unexposed and productive sediments and in deeper waters beneath a very productive water column or following phytoplankton blooms (e.g. in the southern Baltic Sea). The sediments off the upwelling coasts of Chile and Peru are covered by the white sulphur bacterium *Thioploca* (Gallardo 1977; Maier and Gallardo 1984).

The dense mats of white sulphur bacteria create extremely steep gradients of oxygen and sulphide, so that the transition from fully aerobic to anaerobic conditions takes place over a few mm. They also represent zones of high productivity and they harbour a rich biota of heterotrophic microaerophilic bacteria and protists (see below). Finally, the white sulphur bacteria accomplish a rapid regeneration of sulphate close to where it can be consumed by sulphate reducers.

Oxidation of sulphide to sulphate is a proton-producing process. In sediments, the oxic–anoxic boundary layer is therefore characterized by a drop in pH. This is evident even in seawater (which has a relatively high buffering capacity); in the sulphide oxidation zone the pH falls from 7.5–8 to about 7 in marine sediments and then increases again in the sulphidic zone (Fenchel 1969).

In sediments with less intense sulphate reduction, the zone of sulphide oxidation is thicker and consists in part of a mosaic of reduced and microaerobic microhabitats (4.1). In lakes or in marine basins with an anoxic water column, the sulphide–oxygen transition zone is considerably thicker than in sediments since vertical transport and mixing is not due solely to molecular diffusion (Figs 1.6 and 1.7). In the absence of light (absence of competition from phototrophic sulphur bacteria), there is a distinct vertical zone dominated by chemolithotrophic sulphide oxidizers. The chemocline in the water column has been particularly useful in demonstrating and quantifying the rate of reductive CO_2-assimilation of sulphide oxidizers using ^{14}C-labelled bicarbonate (Jørgensen *et al.* 1979; Sorokin 1970, 1972).

5.1.2 *Food chains based on chemolithotrophs*

The fact that chemolithotrophs constitute the base for food chains in aquatic communities first became widely appreciated after the discovery of the biota associated with deep-sea hydrothermal vents (Jannasch 1985; Grassle 1986; Karl 1987), but it has been known for a long time in the case of shallow-water sediments and stratified water columns (Fenchel 1969; Sorokin 1965). (Research which requires millions of dollars and complicated technology invariably enjoys more respect and publicity than results obtained using a more modest fraction of taxpayers' money.)

The oxic–anoxic boundary in sediments or in the water column always harbours a high density of phagotrophic organisms (Figs 5.3 and 5.4). These are predominantly heterotrophic protists (especially different groups of flagellates and ciliates) which graze directly on sulphur bacteria (Fig. 5.1C), which in turn serve as food for predatory protists. The reason why metazoa usually play a modest role in this region is the prevailing low p_{O_2}. The eukaryotic biota is composed of microaerophiles which typically aggregate and thrive at p_{O_2}-levels of 2–10% atm. sat. (Fenchel *et al.* 1989; Fenchel and Finlay 1989; Finlay 1981; Fenchel, unpublished results). Especially in sediments, however, small invertebrates ('meiofauna' including nematodes, turbellaria, gastrotrichs, gnathostomulids, polychaetes and oligochaetes) also play a significant role as consumers of bacteria or other eukaryotes at the oxic–anoxic boundary. The meiofauna, the protists and the bacteria are also utilized by some larger, sediment-dwelling invertebrates (Fenchel 1969; Giere 1992).

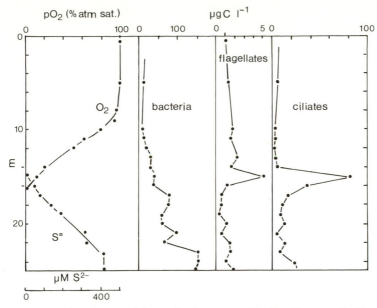

Fig. 5.3 The abundance of bacteria, heterotrophic flagellates and ciliates (expressed as org C/ℓ) in Mariager Fjord, Denmark, October 1987 (data from Fenchel *et al.* 1990).

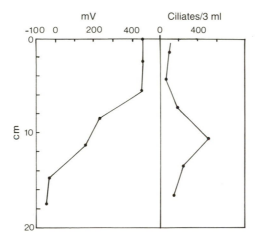

Fig. 5.4 Redox (mV) profile and the vertical quantitative distribution of ciliates in a beach off Helsingør, Denmark, May 1967 (data from Fenchel 1969).

A number of organisms exploit the energy of sulphide (and in a few cases methane) through a symbiotic relationship with chemoautotrophic bacteria. The peculiar ciliate genus *Kentrophoros* does not have a mouth. Instead, the dorsal, non-ciliated side is covered with sulphide-oxidizing bacteria oriented perpendicularly to the host cell surface. The ciliate makes its living by phagocytosis of the symbionts directly from the dorsal side (Fig. 5.5). The *Kentrophoros* species are microaerophiles and the cells accumulate at a p_{O_2} of around 5% atm.sat. (Fenchel and Finlay 1989; Raikov 1971, 1974). *Kentrophoros* is the only protist known to harbour symbiotic sulphur-oxidizing bacteria.

In contrast, there are many representatives of various metazoan groups which depend totally or in part on symbiotic sulphide-oxidizing bacteria. The significance of these symbiotic associations between metazoa and bacteria was first recognized in the case of the hydrothermal vent fauna (Cavanaugh 1983) although the phenomenon occurs in many sediment-dwelling, marine invertebrates from shallower waters. The body surface of nematodes belonging to the Stilbonematinae (*Leptonemella, Catanema*, etc.) is covered with white sulphur bacteria (Fig. 5.5). Apparently, the worms periodically scrape off and eat parts of their bacterial coat. They occur in marine sediments around the oxic–anoxic boundary or they migrate between the sulphidic and the aerobic zone (Jensen 1987a; Ott *et al.* 1982; Schiemer *et al.* 1990). Several species of marine oligochaetes harbour either endosymbiotic or ectosymbiotic sulphur bacteria; in some cases the hosts are totally dependent on their symbionts for nutrition. These forms also live in sediments and show vertical migrations, or they are embedded vertically in the sediment so that part of the body is in the oxic zone (Giere 1981; Giere *et al.* 1984, 1991). Among the bivalves, the genera *Lucina* and *Thyrasira* have gills which are packed with sulphur bacteria. They use their foot to make tubular burrows in the sulphidic part of the sediment, thus gaining access to sulphide while oxygen is supplied from a burrow receiving inhalant water from the sediment surface (Dando *et al.* 1985, 1986; Giere 1985). The shallow-water clam *Solemya* is gutless and entirely dependent on its symbiotic sulphur bacteria (Anderson *et al.* 1987) and a mytilid bivalve has been shown to harbour CH_4-oxidizing bacteria (Childress *et al.* 1986). Bivalves with symbiotic sulphur bacteria (*Bathymodiolus* and *Calyptogena*) are also known from deep-sea vents (Grassle 1986; Southward 1987).

Representatives of the phylum Pogonophora are gutless and their nutrition is based on symbiotic sulphur bacteria (or, in one case, methane oxidizers) maintained in a special organ (the 'trophosome'). Pogonophora are found in marine sediments; they are mostly confined to relatively deep waters (> 300 m) in most seas. The giant worms of hydrothermal vents belong to a special group of pogonophorans, the Vestimentifera (Cavanaugh 1983; Southward 1982, 1987, Schmaljohann and Flügel 1987).

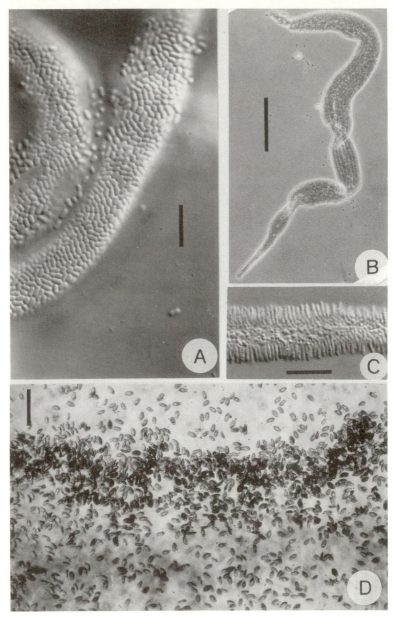

Fig. 5.5 A, the marine benthic nematode *Leptonemella aphanothecae* covered by symbiotic, sulphide-oxidizing bacteria; scale bar 10 μm (photograph by Preben Jensen). B,C, the ciliate *Kentrophoros fasciolata* from marine sands showing its dorsal cover of symbiotic sulphur bacteria; scale bars 100 and 10 μm, respectively. D, the microaerophilic ciliate *Euplotes* sp. in an oxygen gradient; the ciliates accumulate at a p_{O_2} of about 5% atm. sat.

It is conceivable that the original adaptive significance of symbiosis with sulphide oxidizers was, at least in some cases, one of sulphide-detoxification, required for life in sediments close to or within the sulphidic zone and that the exploitation of the bacteria as food is secondary (Vismann 1991*a*; Section 3.5). The function of detoxification has been ascribed to some cases where sulphide oxidizers are found on the surface of sediment dwelling invertebrates, where the bacteria are not consumed by the host (Oeschger and Schmaljohann 1988). Southward (1987) reviewed the symbiosis between invertebrates and symbiotic chemoautotrophic bacteria and the biota of hydrothermal vents have been reviewed by Grassle (1986). Shallow seeps of water containing methane and or sulphide also exist; in these cases the biota are less dependent on chemoautotrophy-based food chains (e.g. Jensen 1986; Jensen *et al.* 1992, Powell *et al.* 1983; P. Dando, personal communication).

Thus, anaerobic habitats support communities of protists and animals in the immediate vicinity and this is based on the flux of reduced end-products of anaerobic metabolism. The chemolithotrophic bacteria which form the basis of the food chains often create steep chemical gradients, resulting in sharp delimitation of the underlying anaerobic world. The energy which maintains these communities is, of course, ultimately derived from the organic material which has been mineralized under anaerobic conditions. The end-products, predominantly methane and sulphide, represent a potential energy source for life only in the presence of oxygen (or light in the case of sulphide). In the case of hydrothermal vents, semi-popular accounts have stressed the fact that life is supported by inorganic compounds which derive from geological, rather than biological processes and it has been implied that such phenomena represent some sort of 'cradle of life'. However, in the absence of light or in the absence of oxygen (which could only be produced by oxygenic photosynthesis elsewhere) sulphide and other reduced compounds do not contain potential energy which can be harvested by organisms. The specialized communities of hydrothermal vents are remarkable in terms of adaptations to high temperatures and to their patchy and ephemeral habitats. With respect to their dependency on chemolithotrophs (either directly as food or as symbionts which are harvested) and with respect to their ability to tolerate and detoxify sulphide, they do not differ from a range of other invertebrates found in most marine sediments. Assuming that there have been episodes during the Phanerozoic when the deeper parts of the oceans were anaerobic (see Section 1.2), it is reasonable to consider the fauna of hydrothermal vents (and the deep-sea fauna in general) as relatively young, a suggestion which is not contradicted by the taxonomic composition of the vent fauna which includes mytilid bivalves and brachyuran crabs.

5.2 The biogeochemical significance of anaerobic life

Although details are unclear, it is obvious that the biogeochemical cycling of the major elements was very different (or 'incomplete' in terms of contemporary conditions) prior to the appearance of atmospheric oxygen. Three major biological elements (C, N and S) all span 8 oxidation levels in biogeochemical cycles ($CH_4 \longleftrightarrow CO_2$ [-4 to $+4$]; $NH_4^+ \longleftrightarrow NO_3^-$[$-3$ to $+5$] and $S^{2-} \longleftrightarrow SO_4^{2-}$ [-2 to $+6$]); see Fig. 5.6. In living organisms, N and S occur in their most reduced form while C usually occurs with an intermediate oxidation level of 0. All three elements are used in dissimilatory metabolism as electron acceptors or as electron donors and all three are important building blocks for all organisms, being assimilated from their surroundings, often after assimilatory reduction. The whole span of oxidation levels of C must have been realized earliest and prior to the advent of atmospheric

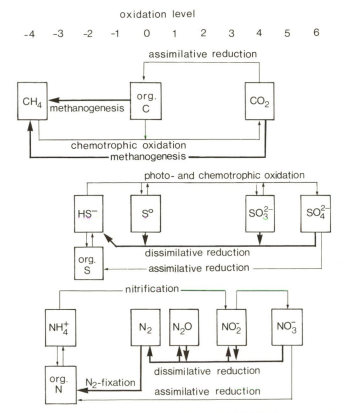

Fig. 5.6 The biogeochemical cycling of C, S and N. The bold arrows represent processes which occur exclusively anaerobically.

oxygen. Sulphate formed on Earth after the advent of phototrophic sulphur bacteria, but probably not earlier. The nitrogen cycle, however, could not have been completed prior to the appearance of oxygen, which is required for the formation of more oxidized forms of N (nitrate, nitrite).

Conversely, there are some processes which occur only in anaerobic habitats and which are of essential importance on a global as well as on local scales in the contemporary biosphere. Quantification of some of these processes on a global scale (e.g. the flux of methane and nitrous oxide from anaerobic habitats) has drawn particular interest due to their effects on atmospheric chemistry and their contribution to possible changes in global climate. Such topics are beyond the scope of this book, but a discussion of the qualitative aspects of anaerobic processes, which are essential to element cycling in the biosphere, is appropriate. For a general treatment of the biogeochemical cycling of elements, see Stumm and Morgan (1970).

There are a number of anaerobic transformations within the carbon cycle which are of biogeochemical significance. These include acid formation in conjunction with fermentation and its effect on the erosion of calcareous rocks and, conversely, the anaerobic deposition of limestone and dolomite due to proton-consuming anaerobic processes. One globally significant aspect is the production of methane which is an exclusively anaerobic process. Most atmospheric methane is believed to derive from swamps, other wetlands and soils, with smaller contributions from symbiotic methanogenesis (notably in ruminants and termites), seeping natural gas (which in some cases also consists of fossil biogenic, rather than thermogenic methane) and from the more recent exploitation of natural gas by man. The methane concentration in the atmosphere has increased during this century; this is believed to be due mainly to methanogenesis in rice paddies and the increased stock of cattle and sheep. The methane which is not oxidized biologically in solution is transformed photochemically in the atmosphere (Bolin *et al.* 1986).

There are two essential processes involving the N-cycle which take place only under anaerobic conditions (Fig. 5.6). One of these is denitrification. Prior to the appearance of O_2, N_2 was a conservative constituent of the atmosphere (excepting the possibility that biological N_2-fixation was important prior to oxygenic photosynthesis). In the presence of oxygen, however, the atmosphere is thermodynamically unstable. Due to the high activation energy of the N_2-molecule, it does not oxidize spontaneously except during electrical discharges (lightning under natural circumstances). Furthermore, biological (and now industrial) N-fixation leads primarily to organic N. After mineralization, it is released as ammonia which is eventually nitrified to nitrate. In the absence of denitrification, atmospheric N_2 would slowly be depleted and accumulate, mainly in the form of dissolved NO_3^-. Thus, the anaerobic process of denitrification is essential for the maintenance of the

N_2-content of the atmosphere (Sillén 1966). Denitrification is also the source of N_2O in the atmosphere (Fenchel and Blackburn 1979).

Reactive N (nitrogen compounds which can be assimilated by plants, algae and bacteria) is often limiting for primary production in the sea. Denitrification represents a sink for reactive N (which is probably balanced by N_2-fixation over some period of time; see below). Denitrification is stimulated by organic material and may thus effect control over primary productivity in the sea. These interactions are, however, still not very well understood (Billen and Lancelot 1988).

The other anaerobic process of the nitrogen cycle is N_2-fixation. The ability to fix atmospheric N_2 is found exclusively among certain prokaryotes. The enzyme nitrogenase is related to hydrogenase and it is inactivated by oxygen. Nitrogen fixation is therefore found primarily among anaerobic, facultative anaerobic or microaerophilic bacteria. Nitrogen-fixing aerobes have special adaptations which protect the enzyme from oxygen. These include excessive energy-uncoupled oxygen consumption and thick walls of mucous which form a diffusion barrier (in *Azotobacter*). Symbiotic nitrogen-fixers live in special host plant organs which provide a low p_{O_2} (e.g. the root nodules of legumes). Heterocysts are found in cyanobacteria: these are specialized cells which occur in some filamentous species. Heterocysts do not have photosystem 2 and hence do not produce oxygen, but cyclic phosphorylation takes place and provides the energy needed for N_2-fixation (Postgate 1982).

The origin and original adaptive significance of nitrogen fixation has been debated. The ability is widespread among anaerobic bacteria (e.g. in clostridia and phototrophs) which may seem strange, since reactive nitrogen (in the form of NH_4^+) is not generally limiting or scarce in anaerobic habitats. (Due to the low energy efficiency of anaerobic heterotrophs, they must dissimilate a relatively larger amount of the organic C of their substrate. Consequently they assimilate a smaller fraction of the available organic N and they will be net-mineralizers of NH_4^+ unless the C/N ratio of the substrate is very high.) A relatively high availability of ammonia has generally been assumed for the pre-oxic conditions on Earth, but this remains speculative. Some fossil filamentous cyanobacteria from the early Precambrian have structures which closely resemble heterocysts and this probably signifies that these organisms were capable of nitrogen fixation. This suggests that reactive N was probably a limiting factor for primary production by the earliest organisms with oxygenic photosynthesis.

The major contribution to the sulphur cycle by anaerobic habitats is dissimilatory sulphate reduction. As previously discussed, the sulphur cycle is largely closed and localized within sediments or around the oxic–anoxic boundary in the water column, since practically all of the sulphide that is produced is re-oxidized in the immediate vicinity of its origin. Some atmospheric sulphur, however, does derive from very shallow marshes and wetlands, but the quantitative importance of this process as compared to other sources (volcanic gases, combustion of fossil fuels) is not well known. Some of the sulphide produced is fossilized in the form of metallic sulphides in sediments. In sulphureta, a large part of the sulphur is temporarily stored in the incompletely re-oxidized S^0; such fossil sulphureta are the source of many exploited sulphur deposits (the remaining being of volcanic origin); see Ivanov (1968).

Phosphorus is an essential element for all life and its availability controls the primary productivity of many ecological systems. The cycling of P differs fundamentally from that of C, N and S in that it does not change its valence (+5) in any biological process and thus plays no role in dissimilatory metabolism, as either an electron acceptor or electron donor. It can be taken up (and released) by bacteria and by phototrophic organisms as ortho-phosphate. In nature, phosphate also occurs in organic forms which are generally available biologically, as polyphosphates, or as biologically inaccessable forms (adsorbed, or as precipitated metallic phosphates). Under aerobic and neutral conditions phosphates tend to precipitate with Ca, Al or Fe. Aquatic habitats therefore act as a global P-sink. Phosphate is continuously transported from the land by rivers and run-off due to the weathering of rocks, and it is returned to the land only over geological time, through mountain building.

Under anaerobic conditions, however, deposited phosphate is in part solubilized due to the formation of metallic sulphides, a process which releases the phosphate. Some phosphate is therefore regenerated from sediments. This mechanism plays an important role in lake eutrophication. As long as the surface layer of the sediment is aerobic, phosphates are retained. But if the sediment surface or the water column becomes anaerobic, large amounts of phosphate are released, resulting in a further increase in biological production (Wetzel 1975).

5.3 The ecological and evolutionary consequences of oxygen in the biosphere

We have previously concluded (Section 3.1) that it is probable that eukaryotes evolved very early in the evolution of life and that the first forms were small, amitochondriate anaerobes which made their living by phagocytosis of bacteria. As in the extant anaerobic fermenters, their energy yield would be only 2–4 moles of ATP per mole of glucose dissimilated and their growth yield only about 25% of that of aerobes (4.2–4.3). In common with extant anaerobic communities, food chains were therefore short and prob-

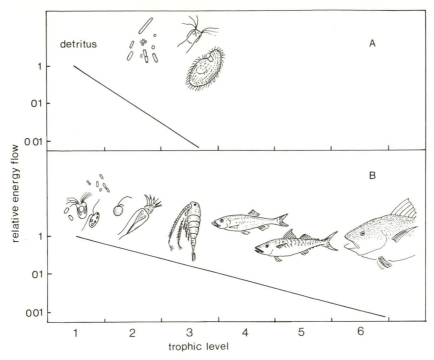

Fig. 5.7 The decrease in energy flow as a function of trophic level in an anaerobic (A) and an aerobic (B) food chain (original).

ably represented by only two trophic levels (bacteria and their eukaryotic predators).

The length of a food chain is determined by the energy loss from one trophic level to the next, a loss which is quantified by growth efficiencies. Assuming a gross growth efficiency of 10% for the anaerobic food chain, then the production of the bacterivorous phagotrophs represents only 1% of the energy input to the system (Fig. 5.7). Under aerobic conditions (assuming a growth efficiency of 40%), the sixth trophic level will still represent a production of 1% of that at the base of the food chain. The maximum number of food chain levels in aerobic communities is often assumed to be about 5–6. Most real systems, however, are too complex to be described in terms of linear food chains. The species populations form a 'food web' in which single species may represent more than one trophic level. However, the basic argument that the number of trophic levels is limited by the growth efficiencies and that anaerobic food chains must, therefore, be much shorter than aerobic ones, is not affected by such complications.

The composition and properties of pre-oxic (viz. earlier than 3.5 billion years ago) biotic communities are, as we have already emphasized, very poorly understood. But it can safely be concluded that, as in extant anaerobic communities, phagotrophs were relatively unimportant and food webs played a small role in structuring biotic communities in comparison with 'horizontal interactions'; viz. syntrophy and competition. The dominating role of predation in the structure, function and diversity of extant aerobic communities is due entirely to the high energetic efficiency of aerobic metabolism.

Animals have evolved to become large and complex, mainly in order to eat larger prey or to avoid being eaten themselves. One explanation why larger anaerobic eukaryotes did not evolve is therefore that food chains had to be very short and the only potential prey were bacteria. If oxygenic photosynthesis had not evolved, eukaryotes would have remained very small, unicellular phagotrophs with a somewhat marginal role in biotic communities.

The advent of oxygen in the atmosphere, of course, led to the evolution of detoxification mechanisms against oxygen-radicals (Section 3.3). More importantly, it allowed for the evolution of aerobic bacteria, and eukaryotes which had acquired mitochondria thus became aerobes. These first aerobes were microaerophiles which could thrive at very low p_{O_2}-levels of perhaps 0.1–1% atm.sat. But the efficiency of their energy metabolism increased by a factor of about ten and this allowed for longer food chains and several trophic levels of unicellular phagotrophs. The acquisition of mitochondria probably took place about 1.5–2 billion years ago according to biochemical data (especially cyctochrome c sequences; see Section 1.2) and resulted in adaptive radiation among the unicellular eukaryotes (see also Section 3.1). This is also supported by geological evidence (Jenkins 1991; see also Section 1.2).

The possibility that there is a causal relation between the appearance and relatively rapid diversification of metazoa and the increasing oxygen level of the atmosphere during the late Proterozoic and the Cambrian has not escaped the notice of several authors (see e.g. Glaessner 1984; Nursall 1959; Runnegar 1991; Schopf and Klein 1992). One special property of metazoa, which has been emphasized in this context, is the neccesary biosynthesis of collagen which is an oxygen dependent process (see Section 3.5). As far as microorganisms are concerned, oxygen had been present in the atmosphere at levels which were sufficient for the bioenergetic and biosynthetic functions of oxygen during a large part of the Precambrian. On the other hand, larger metazoa require p_{O_2}-levels which are typically at least 10–50% of atmospheric saturation (Fig. 5.8). The evolution of metazoa required not only oxic conditions which could satisfy bacteria and small unicellular protists, but a p_{O_2} which approached the present level.

In physiology classes, it is a common exercise to calculate the maximum size of an aerobic (spherical) cell, given the assumption that intracellular

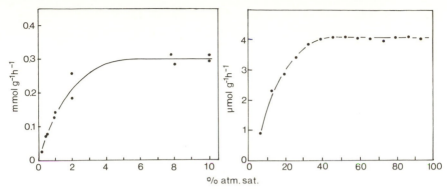

Fig. 5.8 Weight specific respiration rate as a function of pO_2 in the ciliate *Euplotes* sp. (*left*) and the prawn *Palaemon serratus* (*right*). It can be seen that the specific O_2-uptake rate of the ciliate is about 100 times higher than that of the prawn and that the p_{O_2} required for maximum respiratory rate (and for 50% max. respiration = K_m) is about 20 times lower for the ciliate (data on the ciliate from Fenchel *et al.* 1989 and for the prawn, from Taylor 1990).

transport of oxygen takes place by molecular diffusion and that only the center of the cell is anoxic (Krogh 1941). This problem was also considered (in a somewhat different context) in Section 1.2. It was shown that the maximum radius (*r*)of a spherical particle, which consumes oxygen and which is aerobic all the way to its centre, is given by the expression, $r = [6C(0)D/R]^{1/2}$, where $C(0)$ is the external oxygen concentration, D is the diffusion coefficient of O_2 inside the particle (cytoplasm in this case) and R is the volume specific rate of oxygen-uptake (respiration rate). Similar solutions can be found for cylindrical or flattened shapes (Hill 1929; Powell 1989). Given realistic values for R and D and assuming $C(0) = 100\%$ atm.sat., it can be seen that an aerobic spherical protozoon cannot exceed a diameter of 1–2 mm. Some of the assumptions are approximations. Thus, the mitochondria are usually situated close to the periphery of large protozoa. The model also assumes zero-order kinetics for oxygen uptake and this is not realistic when p_{O_2} is very low (Powell 1989). But unicellular organisms do not exceed about 1 mm in diameter (except in some special cases which can be explained; see Fenchel 1987).

Another way to look at this problem is to consider the necessary external p_{O_2} as a function of cell radius. From the above equation we see that $C(0) \propto Rr^2$ (assuming that D is size-invariant). Since $R \propto$ (volume)$^{-0.25} \propto r^{-0.75}$ (Fenchel and Finlay 1983; Hemmingsen 1960), we have that $C(0) \propto r^{1.25}$. In Fig. 5.9, the requirement for external p_{O_2} is expressed as the O_2-tension which supports 50% of the maximum rate of respiration (saturation of oxygen uptake is 2–4 times higher than these half saturation values; cf. Fig. 5.8). In the case of unicellular organisms (data points 11–14), the slope is slightly

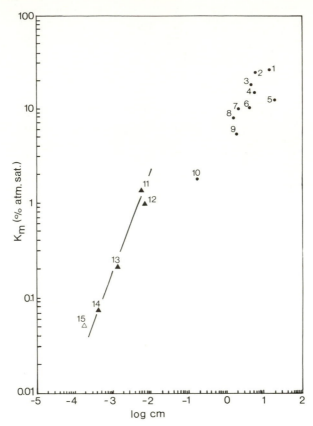

Fig. 5.9 The half saturation p_O for respiration of different organisms/organelles as a function of their linear dimensions. The data points represent the following: crustaceans (1–4,6,8–9), fish (5,7), the oligochaete *Tubifex* (10), ciliates (11,12), the amoeba *Acanthamoeba*, yeast (14) and *Tubifex* mitochondria (15). Data are based on Bradford and Taylor 1982 (1), Bridges and Brand 1980 (4,9), Degn and Kristensen 1981 (10,14,15), Fenchel *et al.* 1989 (12), Hagerman and Szaniawska 1988 (8), Hagerman and Szaniawska 1989 (6), Lloyd *et al.* 1980 (11), McMahon 1988 (2), Taylor 1990 (3), Ultsch *et al.* 1980 (5) and Ultsch *et al.* 1981 (7).

lower (about 1.05) than predicted (1.25). This is probably due to the fact that larger cells tend to deviate more from a spherical shape and the largest diameter has been used as a measure of length on the horizontal axis. The mitochondrion can also be considered to represent an aerobic bacterium. The maximum oxygen uptake of mitochondria is realized at a p_{O_2} of $\approx 0.6\%$ atm. sat. (Dejours 1975).

Data for metazoa are more difficult to interpret. As we have already seen, organisms exceeding a diameter of 1–2 mm cannot rely completely on molecular diffusion for the internal transport of oxygen. Larger animals have a vascular system, respiratory pigments (haemoglobin, haemocyanin) and an allometric increase in their external surface area in the form of gills or lungs. We would therefore expect a discontinuity between the data of unicellular organisms and those for the larger animals (> 1–2 mm). (Very small metazoa without a vascular system will probably behave like unicellular organisms in this respect.) We would also expect greater variability among similarly sized metazoa. Thus, differences in oxygen-binding curves and structural and behavioural differences account for some (size-independent) variation with respect to oxygen uptake as a function of ambient p_{O_2} (Dejours 1975). Still, Fig. 5.9 clearly shows that the external p_{O_2} which is required, increases with increasing body size and eventually approaches atmospheric saturation for the largest aquatic animals.

In large animals, the internal transport of oxygen involves a number of steps which operate at sequentially lower values of p_{O_2}. From the water across the gills to the blood, from the blood to the tissues, from the tissues into the cells and from the cytoplasm to the mitochondria, net oxygen transport is everywhere due to the maintenance of concentration gradients. The mitochondria function at very low oxygen tensions, as did their microaerophilic bacterial ancestor, but the transport of oxygen within large animals from the environment to the mitochondria necessitates a high external p_{O_2}.

In modern seas, the benthic macrofauna disappears when the oxygen tension of the overlying water falls below 7–15% atm. sat. (Elmgren 1978; Tyson and Pearson 1991; Weigelt and Ruhmor 1986). The meiofauna (especially nematodes, turbellaria, gnathostomulids and gastrotrichs) are generally more tolerant of low O_2-tensions and thrive in sediments with a very low p_{O_2} (we exclude here the fact that some meiofauna species may live for periods or perhaps even permanently under anaerobic conditions; see Section 3.5). It is also evident that especially small species (and nematodes with a very small diameter) dominate 'dysaerobic' habitats in sediments (Elmgren 1978; Fenchel and Riedl 1970; Lee and Atkinson 1976; Jensen 1986; Powell 1989; Powell *et al.* 1983; see also Section 3.5).

By analogy, it is reasonable to assume that the first metazoa were microscopic forms similar to extant turbellaria, gastrotrichs, nematodes and rotifers, but some groups developed into progressively larger species as the p_{O_2} increased. The first macroscopic metazoa for which there are recognizable fossil remains, appear in the late Vendian, and towards the end of this period (the Ediacarian) a variety of invertebrate phyla including coelenterates and annelids, and some now extinct groups, had appeared. During the following Cambrian period, further radiation and diversification took place with the appearance of skeletonized forms (Glaessner 1984; Runnegar 1991).

Thus there is an analogy between the evolution of body size as a function of p_{O_2} and the distribution of animals in oxygen gradients (such as in the Black Sea as a function of depth, or along an oxygen gradient caused by

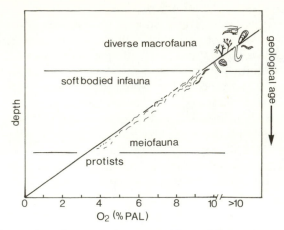

Fig. 5.10 An analogy between the composition of the metazoan benthic fauna along an oxygen gradient and the evolution of metazoa (metazoan size) as a function of geological age during the upper Proterozoic (redrawn from Rhoades and Morse 1971).

anthropogenic eutrophication in a fjord; see Pearson and Rosenberg 1978). This has been illustrated by Rhoads and Morse (1971), as shown in Fig. 5.10.

Another effect of the appearance of atmospheric oxygen was the colonization of land in the early Silurian. The formation of an atmospheric ozone layer which protects the surface of the earth against intense solar ultraviolet radiation is usually quoted, but this may not have required a high atmospheric p_{O_2}. According to Kasting and Donahue (1980) a protective ozone layer would have developed at a p_{O_2} of $\approx 10\%$ of the present level, which was probably attained much earlier. Microbes and a variety of small invertebrates and algae find niches in the water films on soil particles and plant debris. But the evolution of real terrestrial organisms, which are more or less independent of liquid water (vascular plants, arthropods, amniotic vertebrates) required relatively large body sizes which facilitated the development of adaptations against dessication. And as we have argued, large body size required not only aerobic metabolism, but also a p_{O_2} which at least approached that of the extant atmosphere.

Anaerobic and reducing habitats were a prerequisite for the origin of life and all extant life forms have features which reflect this anaerobic origin. Anaerobic habitats and biota have existed continuously since the origin of life during the earliest periods of the history of Earth and anaerobic biota are still an essential component of the biosphere. On the other hand, oxygenic photosynthesis evolved early in the history of life and this process has, since then, dominated the biosphere and at the same time maintained anoxic habitats and biota.

For those who find it a meaningful pastime to speculate on the existence of 'intelligent life' elsewhere in the universe, they might heed the caveat, that the evolution of large and complex forms of life on Earth was only possible due to the advent of atmospheric oxygen and the subsequent evolution of oxidative phosphorylation. This requirement significantly reduces the probability of the evolution of complex life forms on some remote planet.

References

Abram, J.W. and Nedwell, D.B. (1978). Hydrogen as a substrate for methanogenesis and sulfate reduction in anaerobic saltmarsh sediments. *Arch. Microbiol.* 117, 93–97.

Adam, R.D. (1991). The biology of *Giardia* spp. *Microbiol. Rev.* 55, 706–732.

Akin, D.E. and Amos, H.E. (1979). Mode of attack on orchard grass leaf blades by rumen protozoa. *Appl. Environ. Microbiol.* 37, 332–338.

Alldredge, A.L. and Cohen, Y. (1987). Can microscale chemical patches persist in the sea? Microelectrode study of marine snow, faecal pellets. *Science* 235, 689–691.

Aller, R.C. (1983). The importance of the diffusive permeability of animal burrow linings in determining marine sediment chemistry. *J. Mar. Res.* 41, 299–322.

Aller, R.C. and Yingst, J.Y. (1978). Biogeochemistry of tube-dwellings: A study of the sedentary polychaete *Amphitrite ornata* (Leidy). *J. Mar. Res.* 36, 201–254.

Aller, R.C. and Yingst, J.Y. (1985). Effects of the marine deposit-feeders *Heteromastus filiformis* (Polychaeta), *Macoma balthica* (Bivalvia) and *Tellina texana* (Bivalvia) on average sedimentary solute transport, reaction rates, and microbial distributions. *J. Mar. Res.* 43, 615–645.

Alperin, M.S. and Reeburgh, W.S. (1985). Inhibition experiments on anaerobic methane oxidation. *Appl. Environ. Microbiol.* 50, 940–945.

Amesz, J. (1991). Green photosynthetic bacteria and heliobacteria. In: *Variations in Autotrophic Life* (ed. J.M. Shively and L.L. Barton), pp. 99–119. Academic Press, London.

Amesz, J. and Knaff, D.B. (1988). Molecular mechanisms of bacterial photosynthesis. In: Zehnder (1988), pp. 113–178.

Anderson, A.E., Childress, J.J. and Favuzzi, J.A. (1987). Net uptake of CO_2 driven by sulphide and thiosulphate oxidation in the bacterial symbiont-containing clam *Solemya reidi*. *J. exp. Biol.* 133, 1–31.

Andreesen, J.R., Schaupp, A., Neurauter, A., Brown, A. and Ljungdahl, L.G. (1973). Fermentation of glucose, fructose and xylose by *Clostridium thermoaceticum*: effect of metals on growth yield, enzymes, and the synthesis of acetate from CO_2. *J. Bacteriol.* 114, 742–751.

Archer, D.B. and Powell, G.E. (1985). Dependence of the specific growth rate of methanogenic mutualistic cocultures on the methanogen. *Arch. Microbiol.* 141, 133–137.

Arp, A.J., Childress, J.J. and Fisher, C.R.Jr. (1984). Metabolic and bloodgas transport characteristics of the hydrothermal vent bivalve *Calyplogena magnifica. Physiol. Zool.* 57, 648–662.

Augustin, H., Foissner, W. and Adam, H. (1987). Revision of the genera *Acineria, Trimyema* and *Trochiliopsis. Bull. British Mus. (Nat. Hist.) Zool.* Ser. 52, 206–217.

Bagarinao, T. and Vetter, R.D. (1990). Oxidative detoxification of sulphide by mitochondria of the california killifish *Fundulus parvipennis* and the speckled sanddab *Citharichtyes stigmaeus. J. Comp. Physiol. B* 160, 519–527.

Bak, F. and Cypionka, H. (1987). A novel type of energy metabolism involving fermentation of inorganic sulphur compounds. *Nature* 326, 891–892.

Bak, F. and Pfennig, N. (1991). Microbial sulfate reduction in littoral sediment of Lake Constance. *FEMS Microbiol. Ecol.* 85, 31–42.

Baldwin, J.R. and Krebs, H. (1981). The evolution of metabolic cycles. *Nature* 291, 381–382.

Balows, A., Trüber, H.G., Harder, W. and Schleifer, K-H. (ed.) (1992). *The Prokaryotes.* Vols 1–4. Springer-Verlag, New York.

Banage, W.B. (1966). Survival of a swamp nematode (*Dorylaimus* spp.) under anaerobic conditions. *Oikos* 17, 113–120.

Barker, H.A. (1936). On the biochemistry of the methane fermentation. *Arch. Microbiol.* 7, 404–419.

Barr, D.J.S. (1988). How modern systematics relate to the rumen fungi. *BioSystems* 21, 351–356.

Barrett, J. (ed.) (1981). *Biochemistry of Parasitic Helminthes.* Macmillan, London.

Barrett, J. (1984). The anaerobic end-products of helminths. *Parasitology* 88, 179–198.

Barrett, J. (1991). Parasitic helminths. In: Bryant (1991), pp. 146–164.

Bauchop, T. (1989). Biology of gut anaerobic fungi. *BioSystems* 23, 53–64.

Bauchop, T. and Elsden, S.R. (1960). The growth of micro-organisms in relation to their energy supply. *J. Gen. Microbiol.* 23, 457–469.

Bauchop, T. and Mountfort, D.O. (1981). Cellulose fermentation by a rumen anaerobic fungus in both the absence and the presence of rumen methanogens. *Appl. Environ. Microbiol.* 42, 1103–1110.

Baum, R.M. (1984). Superoxide theory of oxygen toxicity is center of heated debate. *Chem. Eng. News* 62, 20–26.

Behm, C. (1991). Fumarate reductase and the evolution of electron transport systems. In: Bryant (1991), pp. 88–108.

Berger, J. and Lynn, D.H. (1992). Hydrogenosome-methanogen assemblages in the echinoid endocommensal plagiopylid ciliates, *Lechriopyla mystax* Lynch, 1930 and *Plagiopyla minuta* Powers, 1933. *J. Protozool.* 34, 4–8.

Bernalier, A., Fonty, G., Bonnemoy, F. and Gouet, P. (1992). Degradation and fermentation of cellulose by the rumen anaerobic fungi in axenic cultures or in association with cellulolytic bacteria. *Curr. Microbiol.* 25, 145–148.

Bernalier, A., Fonty, G., Bonnemoy, F. and Gouet P. (1993). Inhibition of the cellulolytic activity of *Neocallimastix frontalis* by *Ruminococcus flavifaciens. J. Gen. Microbiol.* 139, 873–880.

Berner, R.A. (1963). Electrode studies of hydrogen sulfide in marine sediments. *Geochim. Cosmochim. Acta* 27, 563–575.

Berner, R.A. (ed.) (1971). *Principles of Chemical Sedimentology.* McGraw-Hill, New York.

Bernhard, J.M. (1989). The distribution of benthic Foraminifera with respect to oxygen concentration and organic carbon levels in shallow-water Antarctic sediments. *Limnol. Oceanogr.* 34, 1131–1141.

Bernhard, J.M. and Reimers, C.E. (1991). Benthic foraminiferal populations related to anoxia: Santa Barbara Basin. *Biogeochemistry* 15, 127–149.

Berry, W.B.N. and Wilde, P. (1978). Progressive ventilation of the Oceans — an explanation for the distribution of the lower Palaeozoic black shales. *Am. J. Sci.* 278, 257–275.

Bhat, S., Wallace, R.J. and Ørskov, E.R. (1990). Adhesion of cellulolytic ruminal bacteria to barley straw. *Appl. Environ. Microbiol.* 56, 2698–2703.

Bianchi, M., Marty, D., Teyssié, J.-L. and Fowler, S.W. (1992). Strictly aerobic and anaerobic bacteria associated with sinking particulate matter and zooplankton fecal pellets. *Mar. Ecol. Prog. Ser.* 88, 55–60.

Billen, G. and Lancelot, C. (1988). Modelling benthic nitrogen cycling in temperate coastal ecosystems. In: *Nitrogen Cycling in Coastal Marine Environments* (ed. T.H. Blackburn and J. Sørensen), pp. 341–378. John Wiley, Chichester.

Boaden, P.J.S. (1975). Anaerobiosis, meiofauna and early metazoan evolution. *Zool. Scripta* 4, 21–24.

Boaden, P.J.S. and Platt, H.M. (1971). Daily migration patterns in an intertidal meiobenthic community. *Thalassia Jugoslavica* 7, 1–12.

Bock, E., Koops, H.-P. and Harms, H. (1989). Nitrifying bacteria. In: Schlegel and Bowien (1989), pp. 81–96.

Bolin, B., Döös, B.R., Jäger, J. and Warrick, R.A. (ed.) (1986). *The Greenhouse Effect, Climate Change and Ecosystem.* John Wiley, Chichester.

Boone, D.R. and Bryant, M.P. (1980). Propionate-degrading bacterium, *Syntrophobacter wolinii* sp. nov. gen. nov., from methanogenic ecosystems. *Appl. Environ. Microbiol.* 40, 626–632.

Boone, D.R., Johnson, R.L. and Yitai, L. (1989). Diffusion of the interspecies electron carriers H_2 and formate in methanogenic ecosystems and its implications in the measurement of K_m for H_2 or formate uptake. *Appl. Environ. Microbiol.* 55, 1735–1741.

Bouman, L.A. (1983). A survey of nematodes from the Ems estuary Part II. Species assemblages and association. *Zool. Jahrbücher, Systematik, Ökologie und Geographie der Tiere* 110, 345–376.

Bowden, G.H. and Hamilton, I.R. (1987). Environmental pH as a factor in the competition between strains of the oral streptococci *Streptococcus mutans, S. sanguis,* and '*S. mitior*' growing in continuous culture. *Can. J. Microbiol.* 33, 824–827.

Bradford, S.M. and Taylor, A.C. (1982). The respiration of *Cancer pagurus* under normoxic and hypoxic conditions. *J. Exp. Biol.* 97, 273–288.

Brand, T.V. (1946). Anaerobiosis in invertebrates. *Biodynamical Monographs*, No. 4. Normandy, Missouri.

Brauman, A., Kane, M.D., Labat, M. and Breznak, J.A. (1992). Genesis of acetate and methane by gut bacteria of nutritionally diverse termites. *Science* 257, 1384–1387.

Breznak, J.A. (1982). Intestinal microbiota of termites and other xylophagus insects. *Ann. Rev. Microbiol.* 38, 945–955.

Breznak, J.A. (1984). Biochemical aspects of symbiosis between termites and their intestinal microbiota. In: *Invertebrate Microbial Interactions* (ed. J.M. Anderson, A.D.M. Rayner and D.W.H. Walter), pp. 173–203. Cambridge University Press.

Breznak, J.A. and Kane, M.D. (1990). Microbial H_2/CO_2 acetogenesis in animal guts: nature and nutritional significance. *FEMS Microbiol. Rev.* 87, 309–314.

Bridges, C.R. and Brand, A.R. (1980). Oxygen consumption and oxygen-independence in marine crustaceans. *Mar. Ecol. Prog. Ser.* 2, 133–141.

Brock, T.D. (ed.) (1978). *Thermophilic Microorganisms and Life at High Temperatures.* Springer-Verlag, New York.

Brock, T.D. and Madigan, M.T. (1991). *Biology of Microorganisms* (6th edn). Prentice-Hall, Englewood Cliffs, New Jersey.

Broda, E. (1975). *The Evolution of the Bioenergetic Process.* Pergamon Press, Oxford.

Broers, C.A.M., Stumm, C.K., Vogels, G.D. and Brugerolle, G. (1990). *Psalteriomonas lanterna* gen. nov., sp. nov., a free-living amoeboflagellate isolated from freshwater anaerobic sediments. *Europ. J. Protistol.* 25, 369–380.

Broers, C.A.M., Meijers, H.H.M., Symens, J.C., Stumm, C.K. and Vogels, G.D. (1993). Symbiotic association of *Psalteriomonas vulgaris* n. spec. with *Methanobacterium formicicum. Europ. J. Protistol.* 29, 98–105.

Brugerolle, G. (1991). Cell organization in free-living amitochondriate heterotrophic flagellates. In: Patterson and Larsen (1991), pp. 133–148.

Bryant, C. (ed.) (1991). *Metazoan Life without Oxygen.* Chapman and Hall, London.

Bryant, M.P., Wolin, E.A., Wolin, M.J. and Wolfe, R.S. (1967). *Methanobacillus omelianskii*, a symbiotic association of two species of bacteria. *Arch. Microbiol.* 59, 20–31.

Bryant, M.P., Campbell, L.L., Reddy, C.A. and Crabill, M.R. (1977). Growth of *Desulfovibrio* in lactate or ethanol media low in sulfate in association with H_2-utilizing methanogenic bacteria. *Appl. Environ. Microbiol.* 33, 1162–1169.

Caldwell, D.E. and Tiedje, J.M. (1975a). A morphological study of anaerobic bacteria from the hypolimnia of two Michigan lakes. *Can. J. Microbiol.* 21, 362–376.

Caldwell, D.E. and Tiedje, J.M. (1975b). The structure of anaerobic bacterial communities in the hypolimnia of several Michigan lakes. *Can. J. Microbiol.* 21, 377–385.

Calow, P. (1977). Conversion efficiencies in heterotrophic organisms. *Biol. Rev.* 52, 385–409.

Cappenberg, T.E. and Prins, R.A. (1974). Interrelations between sulfate-reducing and methane-producing bacteria in bottom deposits of a fresh-water lake. III. Experiments with ^{14}C-labelled substrates. *Antonie van Leeuwenhoek J. Microbiol. Serol.* 40, 457–469.

Carlsson, J., Nyberg, G. and Wrethén, J. (1978). Hydrogen peroxide and superoxide radical formation in anaerobic broth media exposed to atmospheric oxygen. *Appl. Env. Microbiol. Ecol.* 36, 223–229.

Castenholz, R.W., Bauld, J. and Jørgensen, B.B. (1990). Anoxygenic microbial mats of hot springs: thermophilic *Chlorobium* spp. *FEMS Microbiol. Ecol.* 74, 325–336.

Caumette, P. (1986). Phototrophic sulfur bacteria and sulfate-reducing bacteria causing red waters in a shallow brackish coastal lagoon (Prévost Lagoon, France). *FEMS Microbiol. Ecol.* 38, 113–124.

Cavalier-Smith, T. (1987a). The origin of eukaryote and archaebacterial cells. *Ann. N.Y. Acad. Sci.* 503, 17–54.

Cavalier-Smith, T. (1987*b*). The simultaneous symbiotic origin of mitochondria, chloroplasts and microbodies. *Ann. N.Y. Acad. Sci.* 503, 55–71.

Cavalier-Smith, T. (1990). Symbiotic origin of peroxisomes. In: *Endocytobiology IV* (ed. P. Nardon, V. Gianiazzi-Pearson, A.M. Grenier, L. Margulis and P.G Soniter), pp. 515–528. Institut Internationale de la Recherche Agronomique, Paris.

Cavalier-Smith, T. (1991*a*). Archamoebas: the ancestral eukaryotes? *BioSystems* 25, 25–38.

Cavalier-Smith, T. (1991*b*). Cell diversification in heterotrophic flagellates. In: Patterson and Larsen (1991), pp. 113–131.

Cavanaugh, C.M. (1983). Symbiotic chemotrophic bacteria in marine invertebrates from sulfide rich habitats. *Nature* 302, 58–61.

Cavedon, K. and Canale-Parola, E. (1992). Physiological interactions between a mesophilic cellulolytic *Clostridium* and a non-cellulolytic bacterium. *FEMS Microbiol. Ecol.* 86, 237–245.

Cavedon, K., Leschine, S.B. and Canale-Parola, E. (1990). Characterization of the extracellular cellulase from a mesophilic *Clostridium* (strain C7). *J. Bacteriol.* 172, 4231–4237.

Cech, T.R. (1985). Self-splicing RNA: Implications for evolution. *Int. Rev. Cytol.* 93, 3–22.

Chang, S., Des Marais, D., Mack, R., Miller, S.C. and Stathearn, G.E. (1983). Prebiotic organic synthesis and the origin of life. In: Schopf (1983), pp. 53–97.

Chapman, D.J. and Schopf, J.W. (1983). Biological and biochemical effects of the development of an aerobic environment. In: Schopf (1983), pp. 302–320.

Childress, J.J., Fisher, C.R., Brooks, J.M., Kennicult, M.C., Bidegare, R. and Anderson, A.E. (1986). A methylotrophic marine molluscan (Bivalvia, Mytilidae) symbiosis: mussels fuelled by gas. *Science* 233, 1306–1308.

Clarke, K.J., Finlay, B.J., Vicente, E., Llorèns, H. and Miracle, M.R. (1993). The complex life-cycle of a polymorphic prokaryote epibiont of the photosynthetic bacterium *Chromatium weissei*. *Arch. Microbiol.* 159, 498–505.

Cohen, Y. and Rosenberg, E. (ed.) (1989). *Microbial Mats.* American Society for Microbiology, Washington, DC.

Cohen, Y., Jørgensen, B.B., Revsbech, N.P. and Poplawski, R. (1986). Adaptation to hydrogen sulfide of oxygenic and anoxygenic photosynthesis among cyanobacteria. *Appl. Environ. Microbiol.* 51, 398–407.

Cole, J.A. and Brown, C.M. (1980). Nitrite reduction to ammonia by ferment circuit in the biological nitrogen cycle. *FEMS Microbiol. Lett.* 7, 65–72.

Coleman, G.S. (1980). Rumen ciliate protozoa. *Adv. Parasitol.* 18,121–173.

Conrad, R., Phelps, T.J. and Zeikus, J.G. (1985). Gas metabolism evidence in support of the juxtaposition of hydrogen-producing and methanogenic bacteria in sewage sludge and lake sediments. *Appl. Environ. Microbiol.* 50, 595–601.

Conrad, R., Schink, B. and Phelps, T.J. (1986). Thermodynamics of H_2-consuming and H_2-producing metabolism in diverse methanogenic environments under in situ conditions. *FEMS Microbiol. Ecol.* 38, 353–360.

Corliss, J.O. (1979). *The Ciliated Protozoa.* Pergamon Press, Oxford.

Corliss, J.O. (1984). The Kingdom Protista and its 45 phyla. *BioSystems* 17, 87–126.

Croome, R.L. and Tyler, P.A. (1984). The microanatomy and ecology of '*Chlorochromatium aggregatum*' in two meromictic lakes in Tasmania. *J. Gen. Microbiol.* 130, 2717–2723.

Crutzen, P.J. (1991) Methane's sinks and sources. *Nature* 350, 380–381.

Czerkawski, J.W. and Cheng, K.-J. (1989). Compartmentation in the rumen. In: Hobson (1989), pp. 361–385.

D'Amelio, E., D'Antoni, Cohen, Y. and Des Marais, D.J. (1989). Comparative functional ultrastructure of two hypersaline submerged cyanobacterial mats: Guerrero Negro, Baja California sur, Mexico and Solar Lake, Sinai, Egypt. In: Cohen and Rosenberg (1989), pp. 97–113.

Dando, P.R., Southward, A.J., Southward, E.C., Terwilliger, N.B. and Terwilliger, R.C. (1985). Sulphur-oxidizing bacteria and haemoglobin in gills of the bivalve mollusc *Myrtea spinifera*. *Mar. Ecol. Prog. Ser.* 23, 85–98.

Dando, P.R., Southward, A.J., Southward, E.C. and Barret, R.L. (1986). Possible energy sources for chemoautotrophic prokaryotes symbiotic with invertebrates from a Norwegian fjord. *Ophelia* 26, 135–150.

Dando, P.R., Fenchel, T., Jensen, P., O'Hara, S.C.M. and Schuster, U. (1993). The ecology of gassy, organic rich sediment off a sandy beach on the Kattegat coast of Denmark. *Mar. Ecol. Prog. Ser.* 100, 265–271.

Danson, M.J. and Hough, D.W. (1992). The enzymology of archaebacterial pathways of central metabolism. *Biochem. Soc. Symp.* 58, 7–21.

Davison, W. and Finlay, B.J. (1986). Ferrous iron and phototrophy as alternative sinks for sulphide in the anoxic hypolimnia of two adjacent lakes. *J. Ecol.* 74, 663–673.

Dayhoff, M.O. and Schwartz, R.M. (1981). Evidence on the origin of eukaryotic mitochondria from protein and nucleic acid sequences. *Ann. N.Y. Acad. Sci.* 361, 92–104.

de Duve, C. (1969). Evolution of the peroxysome. *Ann. N.Y. Acad.Sci.* 168, 369–381.

Degn, H. and Kristensen, B. (1981). Low sensitivity of *Tubifex* spp. respiration to hydrogen sulfide and other inhibition. *Comp. Biochem. Physiol.* 69 B, 809–817.

Dejours, P. (ed.) (1975). *Principles of Comparative Respiratory Physiology.* North-Holland, Amsterdam.

Delihas, N. and Fox, G.E. (1987). Origins of the plant chloroplasts and mitochondria based on comparisons of 5S ribosomal RNAs. *N.Y. Acad. Sci.* 503, 92–102.

Deuser, W.G. (1971). Organic-carbon budget of the Black Sea. *Deep-Sea Res.* 18, 995–1005.

Deuser, W.G. (1975). Reducing environments. In: *Chemical Oceanography* (2nd edn) (ed. J.P. Riley and G. Skirrow), Vol. 3, pp. 1–37. Academic Press, London.

de Zwaan, A. (1991). Molluscs. In: Bryant (1991), pp. 186–217.

de Zwaan, A. and Putzer, V. (1985). Metabolic adaptations of intertidal invertebrates to environmental hypoxia (a comparison of environmental anoxia to exercise anoxia). In: *Physiological Adaptations of Marine Animals* (ed. M.S. Laverack), pp. 33–62. Society for Experimental Biology, Cambridge.

Doddema, H.J., van der Drift, C., Vogels, G.D. and Veenhuis, M. (1979). Chemiosmotic coupling in *Methanobacterium thermoautotrophicum*: hydrogen-dependent adenosine 5'-triphosphate synthesis by subcellular particles. *J. Bacteriol.* 140, 1081–1089.

Dolfing, J. (1988). Acetogenesis. In: Zehnder (1988), pp. 417–468.

Drews, G. and Imhoff, J.F. (1991). Phototrophic purple bacteria. In: *Variations in Autotrophic Life* (ed. J.M. Shively and L.L. Barton), pp. 51–97. Academic Press, London.

Duffy, J.E. and Tyler, S. (1984). Quantitative differences in mitochondrial ultrastructure of a thiobiotic and an oxybiotic turbellarian. *Mar. Biol.* 83, 95–102.

Dworkin, M. (1992). Prokaryotic diversity. In: Balows et al. (1992), pp. 48–74.

Dye, A.H. (1983). Composition and seasonal fluctuations of meiofauna in a southern African mangrove estuary. *Mar. Biol.* 73, 165–170.

Eichel, H.J. and Rem, L.T. (1962). Respiratory enzyme studies in *Tetrahymena pyriformis*. V. Some properties of an L-lactic oxidase. *J. Biol. Chem.* 237, 940–945.

Eigen, M. (1992). *Steps towards Life.* Oxford University Press.

Ellis, J.E., Williams, R., Cole, D., Cammack, R. and Lloyd, D. (1993). Electron transport components of the parasitic protozoon *Giardia lamblia*. *FEBS Letters* 325, 196–200.

Elmgren, R. (1978). Structure and dynamics of Baltic benthos communities, with particular reference to the relationships between macro- and meiofauna. *Kieler Meeresforsch. Sonderheft.* 4, 1–22.

Embley, T.M. and Finlay, B.J. (1994). The use of small subunit rRNA sequences to unravel the relationships between anaerobic ciliates and their methanogen endosymbionts. *Microbiology* 140, 225–235.

Embley, T.M., Finlay, B.J. and Brown, S. (1992a). RNA sequence analysis shows that the symbionts in the ciliate *Metopus contortus* are polymorphs of a single methanogen species. *FEMS Microbiol. Lett.* 97, 57–62.

Embley, T.M., Finlay, B.J., Thomas, R.H. and Dyal, P.L. (1992b). The use of rDNA sequences and fluorescent probes to investigate the phylogenetic positions of the anaerobic ciliate *Metopus palaeformis* and its archaeobacterial endosymbiont. *J. Gen. Microbiol.* 138, 1479–1487.

Emery, K.O. and Hunt, J.M. (ed.) (1974). *The Black Sea.* Memoirs of the American Association for Petrol. Geol., Tulsa.

Erwin, D.H. (1994). The Permo-Triassic extinction. *Nature,* 367, 231–236.

Esteban, G. and Finlay, B.J. A new genus of anaerobic scuticociliate with endosymbiotic methanogens and ectobiotic bacteria. *Archiv. Protistenk.* (In press).

Esteban, G., Guhl, B.E., Clarke, K.J., Embley, T.M. and Finlay, B.J. (1993a). *Cyclidium porcatum* n. sp.: a free-living anaerobic scuticociliate containing a stable complex of hydrogenosomes, eubacteria and archaeobacteria. *Europ. J. Protistol.* 29, 262–270.

Esteban, G., Finlay, B.J. and Embley, T.M. (1993b). New species double the diversity of anaerobic ciliates in a Spanish lake. *FEMS Microbiol. Lett.* 109, 93–100.

Famme, P. and Knudsen, J. (1984). Total heat balance study of anaerobiosis in *Tubifex tubifex* (Müller). *J. Comp. Physiol.* B 154, 587–591.

Famme, P. and Knudsen, J. (1985). Anoxic survival, growth and reproduction by the freshwater annelid *Tubifex* sp., demonstrated using a new simple anoxic chemostat. *Comp. Biochem. Physiol.* 81 A, 251–253.

Farmer, M.A. (1993). Ultrastructure of *Ditrichomonas honigbergii* N.G., N. Sp. (Parabasalia) and its relationship to amitochondrial protists. *J. Euk. Microbiol.* 40, 619–626.

Fauque, G., Legall, J. and Barton, L.L. (1991). Sulfate-reducing and sulfur-reducing bacteria. In: *Variations in Autotrophic Life* (ed. J.M. Shively and L.L. Barton), pp. 271–337. Academic Press, London.

Fauré-Fremiet, E. (1951). The marine sand-dwelling ciliates of Cape Cod. *Biol. Bull. Mar. Biol. Lab. Woods Hole* 100, 59–70.

Fenchel, T. (1968). The ecology of marine microbenthos. II. The food of marine benthic ciliates. *Ophelia* 5, 73–121.

Fenchel, T. (1969). The ecology of marine microbenthos. IV. Structure and function of the benthic ecosystem. *Ophelia* 6, 1–182.

Fenchel, T. (1978). The ecology of micro- and meiobenthos. *Ann. Rev. Ecol. Syst.* 9, 99–121.

Fenchel, T. (1987). *Ecology of Protozoa.* Science Tech Publishers, Madison/Springer-Verlag, Berlin.

Fenchel, T. (1993). Methanogenesis in marine shallow water sediments: the quantitative role of anaerobic protozoa with endosymbiotic methanogenic bacteria. *Ophelia* 37, 67–82.

Fenchel, T. and Bernard, C. (1993*a*). A purple protist. *Nature* 362,300.

Fenchel, T. and Bernard, C. (1993*b*). Endosymbiotic purple non-sulphur bacteria in an anaerobic ciliated protozoon. *FEMS Microbiol Lett.* 110, 21–25.

Fenchel, T. and Blackburn, T.H. (1979). *Bacteria and Mineral Cycling.* Academic Press, London

Fenchel, T. and Finlay, B.J. (1983). Respiration rates in heterotrophic, free-living protozoa. *Microbiol. Ecol.* 9, 99–120.

Fenchel, T. and Finlay, B.J. (1984). Geotaxis in the ciliated protozoon *Loxodes. J. Exp. Biol.* 110, 17–33.

Fenchel, T. and Finlay, B.J. (1986*a*). The structure and function of Müller vesicles in loxodid ciliates. *J. Protozool.* 3, 69–76.

Fenchel, T. and Finlay, B.J. (1986*b*). Photobehavior of the ciliated protozoon *Loxodes*: taxic, transient and kinetic responses in the presence and absence of oxygen. *J. Protozool.* 33, 139–145.

Fenchel, T. and Finlay, B.J. (1989). *Kentrophoros*: a mouthless ciliate with a symbiotic kitchen garden. *Ophelia* 30, 75–93.

Fenchel, T. and Finlay, B.J. (1990*a*). Anaerobic freeliving protozoa: growth efficiencies and the structure of anaerobic communities. *FEMS Microbiol. Ecol.* 74, 269–276.

Fenchel, T. and Finlay, B.J. (1990*b*). Oxygen toxicity, respiration and behavioural responses to oxygen in freeliving anaerobic ciliates. *J. Gen. Microbiol.* 136, 1953–1959.

Fenchel, T. and Finlay, B.J. (1991*a*). The biology of freeliving anaerobic ciliates. *Europ. J. Protistol.* 26, 201–215.

Fenchel, T. and Finlay, B.J. (1991*b*). Endosymbiont methanogenic bacteria in anaerobic ciliates: significance for the growth efficiency of the host. *J. Protozool.* 38, 18–22.

Fenchel, T. and Finlay, B.J. (1991*c*). Synchronous division of an endosymbiotic methanogenic bacterium in the anaerobic ciliate *Plagiopyla frontata* Kahl. *J. Protozool.* 38, 22–28.

Fenchel, T. and Finlay, B.J. (1992). Production of methane and hydrogen by anaerobic ciliates containing symbiotic methanogens. *Arch. Microbiol.* 157, 475–480.

Fenchel, T. and Ramsing, N.B. (1992). Identification of sulphate-reducing ectosymbiotic bacteria from anaerobic ciliates using 16S rRNA binding oligonucleotide probes. *Arch. Microbiol.* 158, 394–397.

Fenchel, T. and Riedl, R.J. (1970). The sulfide system: a new biotic community underneath the oxidized layer of marine sand bottoms. *Mar. Biol.* 7, 255–268.

Fenchel, T. and Straarup, B.J. (1971). Vertical distribution of photosynthetic pigments and the penetration of light in marine sediments. *Oikos* 22, 172–182.

Fenchel, T., Perry, T. and Thane, A. (1977). Anaerobiosis and symbiosis with bacteria in free-living ciliates. *J. Protozool.* 24, 154–163.

Fenchel, T., McRoy, C.P., Ogden, J.C., Parker, P. and Rainey, W.E. (1979). Symbiotic cellulose degradation in green turtles, *Chelonia mydas* L. *Appl. Environ. Microbiol.* 37, 348–350.

Fenchel, T., Finlay, B.J. and Gianni, A. (1989). Microaerophilic behaviour in ciliates: responses to oxygen tension in an *Euplotes* species (Hypotrichida). *Arch. Protistenk.* 137, 317–330.

Fenchel, T., Kristensen, L.D. and Rasmussen, L. (1990). Water column anoxia: vertical zonation of planktonic protozoa. *Mar. Ecol. Prog. Ser.* 62, 1–10.

Fiebig, K. and Gottschalk, G. (1983). Methanogenesis from choline by a coculture of *Desulfovibrio* sp. and *Methanosarcina barkeri. Appl. Environ. Microbiol.* 45, 161–168.

Finlay, B.J. (1980). Temporal and vertical distribution of ciliophoran communities in the benthos of a small eutrophic loch with particular reference to the redox profile. *Freshwater Biol.* 10, 15–34.

Finlay, B.J. (1981). Oxygen availability and seasonal migrations of ciliated protozoa in a freshwater lake. *J. Gen. Microbiol.* 123, 173–178.

Finlay, B.J. (1982). Effects of seasonal anoxia on the community of benthic ciliated protozoa in a productive lake. *Arch. Protistenk.* 125, 215–222.

Finlay, B.J. (1985). Nitrate respiration by protozoa (*Loxodes* spp.) in the hypolimnetic nitrite maximum of a freshwater pond. *Freshwater Biol.* 15, 333–346.

Finlay, B.J. (1990). Physiological ecology of freeliving protozoa. *Adv. Microb. Ecol.* 11, 1–35.

Finlay, B.J. and Berninger, U.-G. (1984). Coexistence of congeneric ciliates (Karyorelictida: *Loxodes*) in relation to food resources in two freshwater lakes. *J. Anim. Ecol.* 53, 929–943.

Finlay, B.J. and Fenchel, T. (1986). Photosensitivity in the ciliated protozoon *Loxodes*: pigment granules, absorption and action spectra, blue light perception and ecological significance. *J. Protozool.* 33, 534–542.

Finlay, B.J. and Fenchel, T. (1989). Hydrogenosomes in some anaerobic protozoa resemble mitochondria. *FEMS Microbiol. Lett.* 65, 311–314.

Finlay, B.J. and Fenchel,T. (1991a). An anaerobic protozoon, with symbiotic methanogens, living in municipal landfill material. *FEMS Microbiol. Ecol.* 85, 169–180.

Finlay, B.J. and Fenchel,T. (1991b). Polymorphic bacterial symbionts in the anaerobic ciliated protozoon *Metopus. FEMS Microbiol. Lett.* 79, 187–190.

Finlay, B.J. and Fenchel, T. (1992). An anaerobic ciliate as a natural chemostat for the growth of endosymbiotic methanogens. *Europ. J. Protistol.* 28, 127–137.

Finlay, B.J. and Fenchel, T. (1993). Methanogens and other bacteria as symbionts of free-living anaerobic ciliates. *Symbiosis* 14, 375–390.

Finlay, B.J., Span, A.S.W. and Harman, J.M.P. (1983). Nitrate respiration in primitive eukaryotes. *Nature* 303, 333–336.

Finlay, B.J., Fenchel, T. and Gardener, S. (1986). Oxygen perception and O_2 toxicity in the freshwater ciliated protozoon *Loxodes. J. Protozool.* 33, 157–165.

Finlay, B.J., Berninger, U.-G., Clarke, K.J., Cowling, A.J., Hindle, R.M. and Rogerson, A. (1988). On the abundance and distribution of protozoa and their food in a productive freshwater pond. *Europ. J. Protistol.* 23, 205–217.

Finlay, B.J., Clarke, K.J., Vicente, E. and Miracle, M.R. (1991). Anaerobic ciliates from sulphide-rich solution lake in Spain. *Europ. J. Protistol.* 27, 148–159.

Finlay, B.J., Embley, T.M. and Fenchel, T. (1993). A new polymorphic methanogen, closely related to *Methanocorpusculum parvum* living in stable symbiosis within the anaerobic ciliate *Trimyema* sp. *J. Gen. Microbiol.* 139, 371–378.

Finlay, B.J., Esteban, G., Clarke, K.J., Williams, A.G., Embley, T.M. and Hirt, R.P. (1994). Some rumen ciliates have endosymbiotic methanogens. *FEMS Microbiol. Lett.* 117, 157–162.

Ford, T.E. (ed.) (1993). *Aquatic Microbiology.* Blackwell Scientific Publications, Oxford.

Forterre, P., Benachenhou-Lahfa, N., Confalonieri, F., Duguet, M., Elie, C. and Labedan, B. (1993). The nature of the last universal ancestor and the root of the tree of life, still open questions. *BioSystems* 28, 15–32.

Froelich, P.N., Klinkhauer, G.P. and Bender, M.L. (1979). Early diagenesis of organic matter in pelagic sediments of the eastern equatorial Atlantic: suboxic diagenesis. *Geochim. Cosmochim. Acta* 43, 1075–1090.

Fuchs, G., Ecker, A. and Strauss, G. (1992). Bioenergetics and autotrophic carbon metabolism of chemolithotrophic archaebacteria. *Biochem. Soc. Symp.* 58, 23–39.

Fujimoto, D. and Prockop, D.J. (1969). Protocollagen proteine hydroxylase from *Ascaris lumbricoides. J. Biol. Chem.* 244, 205–210.

Gallardo, V.A. (1977). Large benthic microbial communities in sulphide biota under Peru-Chile subsurface countercurrent. *Nature* 268, 331–332.

Gelhaye, E., Benoit, L., Petitdemange, H. and Gay, R. (1993). Adhesive properties of five mesophilic, cellulolytic *Clostridia* isolated from the same biotope. *FEMS Microbiol Ecol.* 102, 67–73.

Gerritse, J. and Gottschal, J.C. (1993). Two-membered mixed cultures of methanogenic and aerobic bacteria in O_2-limited chemostats. *J. Gen Microbiol.* 139, 1853–1860.

Gest, H. (1980). The evolution of biological energy-transducing systems. *FEMS Microbiol. Lett.* 7, 73–77.

Gest, H. (1987). Evolutionary roots of the citric acid cycle in prokaryotes. *Biochem. Soc.Symp.* 54, 3–16.

Gest, H. and Favinger, J.L. (1983). *Heliobacterium chlorum*, an anoxygenic brownish-green bacterium containing a 'new' form of bacteriochlorophyll. *Arch. Microbiol.* 136, 11–16.

Gest, H. and Schopf, J.W. (1983). Biochemical evolution of anaerobic energy conversion: the transition from fermentation to anoxygenic photosynthesis. In: Schopf (1983), pp. 135–148.

Ghiorse, W.C. (1988). Microbial reduction of manganese and iron. In: Zehnder (1988), pp. 305–321.

Ghiorse, W.C. (1989). Manganese and iron as physiological electron donors and acceptors in aerobic-anaerobic transition zones. In: Cohen and Rosenberg (1989), pp. 163–169.

Giere, O. (1981). The gutless marine oligochaete *Phallodrilus leukodermatus*. Structural studies on an aberrant tubificid associated with bacteria. *Mar. Ecol. Prog. Ser.* 5, 353–357.

Giere, O. (1985). Structure and position of bacterial endosymbionts in the gill filaments of *Lucinidae* from Bermuda (Mollusca, Bivalvia). *Zoomorphology* 105, 296–301.

Giere, O. (1992). Benthic life in sulfidic zones of the sea—ecological and structural adaptations to a toxic environment. *Verh. Dtsch. Zool. Ges.* 85, 77–93.

Giere, O., Felbeck, H., Dawson, R. and Liebezeit, G. (1984). The gutless oligochaete *Phallodrilus leukodermatus* Giere, a tubificid of structural, ecological and physiological relevance. *Hydrobiologia* 115, 83–89.

Giere, O., Conway, N.M., Gashoch, G. and Schmidt, C. (1991). 'Regulation' of gutless annelid ecology by endosymbiotic bacteria. *Mar. Ecol. Prog. Ser.* 68, 217–299.

Gijzen, H.J., Broers, C.A.M., Barughare, M. and Stumm, C.K. (1991). Methanogenic bacteria as endosymbionts of the ciliate *Nyctotherus ovalis* in the cockroach hindgut. *Appl. Environ, Microbiol.* 57, 1630–1634.

Glaessner, M.F. (ed). (1984). *The Dawn of Animal Life*. Cambridge University Press.

Goosen, N.K., Horemans, A.M.C., Hillebrand, S.J.W., Stumm, C.K. and Vogels, G.D. (1988). Cultivation of the sapropelic ciliate *Plagiopyla nasuta* Stein and isolation of the endosymbiont *Methanobacterium formicicum*. *Arch. Microbiol.* 150, 165–170.

Goosen, N.K., van der Drift, C., Stumm, C.K. and Vogels, G.D. (1990). End products of metabolism in the anaerobic ciliate *Trimyema compressum*. *FEMS Microbiol. Lett.* 69, 171–176.

Gorrell, T.E., Yarlett, N. and Müller, M. (1984). Isolation and characterization of *Trichomonas vaginalis* ferredoxin. *Carlsberg Res. Commun.* 449, 259–268.

Gottschal, J.C. and Prins, R.A. (1991). Thermophiles: a life at elevated temperatures. *TREE* 6, 157–162.

Gottschalk, G. and Peinemann, S. (1992). In: Balows *et al.* (1992), Vol. I. pp. 300–311.

Grassé, P.P. (ed.) (1952). *Traité de zoologie*. Vol. I. Masson, Paris.

Grasshoff, K. (1975). The hydrochemistry of landlocked basins and fjords. In: *Chemical Oceanography* (2nd edn) (ed. J.P. Riley and G. Skirrow), Vol. 2, pp. 455–597. Academic Press, London.

Grassle, J.F. (1986). The ecology of deep sea hydrothermal vent communities. *Adv. Mar. Biol. Ecol.* 23, 301–362.

Guerrero, R., Montesinos, E., Pedrós-Alió, C., Esteve, I., Mas, J., Gemerden, H.van, Hofman, P.A.G. and Bakker, J.F. (1985). Phototrophic sulfur bacteria in two Spanish lakes: vertical distribution and limiting factors. *Limnol. Oceanogr.* 30, 919–931.

Guerrero, R., Pedrós-Alió, C., Esteve, I., Mas, J., Chase, D. and Margulis, L. (1986). Predatory prokaryotes: predation and primary consumption evolved in bacteria. *Proc. Natl. Acad. Sci. USA* 83, 2138–2142.

Guhl, B.E. and Finlay, B.J. (1993). Anaerobic predatory ciliates track seasonal migrations of planktonic photosynthetic bacteria. *FEMS Microbiol. Lett.* 107, 313–316.

Guhl, B.E., Finlay, B.J. and Schink, B. (1994). The seasonal development of hypolimnetic ciliate communities in a eutrophic lake. *FEMS Microbiol. Ecol.* 14, 293–306.

Hadas, O., Pinjas, R. and Wynne, D. (1992). Nitrate reductase activity, ammonium regeneration, and orthophosphate uptake in protozoa isolated from Lake Kinneret, Israel. *Microbiol. Ecol.* 23, 107–115.

Hagermann, L. and Szaniawska, A. (1988). Respiration, ventilation and circulation under hypoxia in the glacial relict *Saduria (Mesidotea) entomon. Mar. Ecol. Prog. Ser.* 47, 55–63.

Hagermann, L. and Szaniawska, A. (1989). Respiration during hypoxia of the shrimps *Crangon crangon* and *Palaemon adspersus* from Brackish water. *Proceedings of the 21st European Marine Biology Symposium, Gdansk 1986, Polish Acad. Sci.*

Haldane, J.B.S. (1928). The origin of life. *Rationalist Annual* 148, 3–10.

Hansen, T.A. (1988). Physiology of sulphate-reducing bacteria. *Microbiol. Sci.* 5, 81–84.

Hansen, M.H., Ingvorsen, K. and Jørgensen, B.B. (1978). Mechanisms of hydrogen sulfide release from coastal marine sediments to the atmosphere. *Limnol. Oceanogr.* 23, 68–76.

Harmsen, J.M., Wullings, B., Akkermans, A.D.L., Ludwig, W. and Stams, A.J.M. (1993). Phylogenetic analyses of *Syntrophobacter wolinii* reveals a relationship with sulfate-reducing bacteria. *Arch. Microbiol.* 160, 238–240.

Hemmingsen, A.M. (1960). Energy metabolism as related to body size and respiratory surfaces and its evolution. *Rep. Steno Mem. Hosp. Copenhagen* 9, 1–110.

Herwig, R.P. and Staley, J.T. (1986). Anaerobic bacteria from the digestive tract of North Atlantic fin whales (*Balaenoptera physalus*). *FEMS Microbiol. Ecol.* 38, 361– 371.

Hill, A.V. (1929). The diffusion of oxygen and lactic acid through tissues. *Proc. R. Soc. (London) B* 104, 39–96.

Hillman, K., Lloyd, D. and Williams, A.G. (1985). Use of a portable spectrometer for the measurement of dissolved gas concentration in bovine rumen liquid *in situ. Current Microbiol.* 12, 335–340.

Hillman, K., Lloyd, D. and Williams, A.G. (1988). Interactions between the methanogen *Methanosarcina barkeri* and rumen holotrich ciliate protozoa. *Lett. Appl. Microbiol.* 7, 49–53.

Hobson, P.N. (ed.) (1989). *The Rumen Microbial Ecosystem.* Elsevier, London.

Hochachka P.W. (ed.) (1980). *Living Without Oxygen.* Harvard University Press, Cambridge Mass.

Hochachka, P.W. (1986). Defence strategies against hypoxia and hypothermia. *Science* 231, 234–241.

Hogan, K.B., Hoffman, J.S. and Thompson, A.M. (1991). Methane on the greenhouse agenda. *Nature* 354, 181–182.

Holland, H.D. (ed.) (1984). *The Chemical Evolution of the Atmosphere and Oceans.* Princeton University Press, Princeton N.J.

Holler, S. and Pfennig, N. (1991). Fermentation products of the anaerobic ciliate *Trimyema compressum* in monoxenic cultures. *Arch. Microbiol.* 156, 327–334.

Honigberg, B.M. (1970). Protozoa associated with termites and their role in digestion. In: *Biology of Termites* (ed. K. Krishna and F.M. Wasner), Vol. 2. Academic Press, London.

Howarth, R.W. (1993). Microbial processes in salt-marsh sediments. In: Ford (1993), pp. 239–259.

Hrdý, I. and Mertens, E. (1993). Purification and partial characterization of malate dehydrogenase (decarboxylating) from *Tritrichomonas foetus* hydrogenosomes. *Parasitology* 107, 379–385.

Hungate, R.E. (1955). Mutualistic intestinal protozoa. In: *Biochemistry and Physiology of Protozoa* (ed. S.H. Hutner and A. Lwoff), Vol. II, pp. 159–199. Academic Press, New York.

Hungate, R.E. (ed.) (1966). *The Rumen and its Microbes.* Academic Press, New York.

Hungate, R.E. (1969). A roll tube method for the cultivation of strict anaerobes. In: *Methods in Microbiology* (ed. J.R. Norris and D.W. Ribbons), Vol. 3 B, pp. 117–132. Academic Press, New York.

Hungate, R.E. (1975). The rumen microbial system. *Ann. Rev. Ecol. Syst.* 6, 39–66.

Hungate, R.E. (1978). The rumen protozoa. In: *Parasitic Protozoa* (ed. J.P. Kreier), Vol. II, pp. 655–695. Academic Press, New York.

Hutchinson, G.E., Deevey, E.S. and Wollack, A. (1939). The oxidation-reduction potentials of lake waters and their ecological significance. *Proc. Nat. Acad. Sci. USA* 25, 87–90.

Hylleberg, J. and Henriksen, K. (1980). The central role of bioturbation in sediment mineralization and element re-cycling. *Ophelia* Suppl. 1, 1–16.

Ianotti, E.L., Kafkewitz, P., Wolin, M.J. and Bryant, M.P. (1973). Glucose fermentation products of *Ruminococcus albus* grown in continuous culture with *Vibrio succinogenes*: changes caused by interspecies transfer of H_2. *J. Bacteriol.* 114, 1231–1240.

Indrebø, G., Pengerud, B. and Dundas, I. (1979). Microbial activities in a permanently stratified estuary. I. Primary production and sulfate reduction. *Mar. Biol.* 51, 295–304.

Ingvorsen, K. and Jørgensen, B.B. (1984). Kinetics of sulfate uptake by freshwater and marine species of *Desulfovibrio. Arch. Microbiol.* 139, 61–66.

Ingvorsen, K., Zeikus, J.G. and Brock, T.D. (1981). Dynamics of bacterial sulfate reduction in a eutrophic lake. *Appl. Environ. Microbiol.* 42, 1029–1036.

Inoue, T. and Orgel, L.E. (1983). A non-enzymatic RNA polymerase model. *Science* 219, 859–862.

Ivanov, M.V. (ed). (1968). *Microbial Processes in the Formation of Sulfur Deposits.* Israel Program for Scientific Translations, Jerusalem.

Iversen, N. and Blackburn, T.H. (1981). Seasonal rates of methane oxidation in anoxic marine sediments. *Appl. Environ. Microbiol.* 41, 1295–1300.

Jankowski, A.W. (1964). Morphology and Evolution of Ciliophora. III. Diagnoses and phylogenesis of 53 sapropelebionts mainly of the order heterotrichida. *Arch. Protistenk.* 107, 185–294.

Jannasch, H.W. (1985). The chemosynthetic support of life and the microbial diversity at deep sea hydrothermal vents. *Proc. Roy. Soc. (London)* B 225, 277–297.

Jannasch, H.W. and Mottl, M.J. (1985). Geomicrobiology of deep-sea hydrothermal vents. *Science* 229, 717–725.

Jenkins, R.J.F. (1991). The early environment. In: Bryant (1991), pp. 38–64.

Jensen, P. (1983). Meiofaunal abundance and vertical zonation in a sublittoral soft bottom, with a test of the Haps corer. *Mar Biol.* 74, 319–326.

Jensen, P. (1986). Nematode fauna in the sulphide-rich brine seep and adjacent bottoms of the East Flower Garden, NW Gulf of Mexico. IV. Ecological aspects. *Mar. Biol.* 92, 489–503.

Jensen, P. (1987*a*). Feeding ecology of free-living aquatic nematodes. *Mar. Ecol. Prog. Ser.* 35, 187–196.

Jensen, P. (1987*b*). Differences in microhabitat, abundance, biomass and bodysize between oxybiotic and thiobiotic free-living marine nematodes. *Oecologia* (Berlin) 71, 564–567.

Jensen, P. (1992). *Cerianthus vogti* Danielssen, 1890 (Anthozoa: Ceriantharia). A species inhabiting an extended tube system deeply buried in deep-sea sediments off Norway. *Sarsia* 77, 75–80.

Jensen, P., Aagaard, I., Burke, R.A., Dando, P.R., Jørgensen, N.D., Kuijpers, A., Laier, T., O'Hara, S.C.M. and Schmaljohann, R. (1992). 'Bubbling reefs' in the Kattegat: Submarine landscapes of carbonate-cemented rocks support a diverse ecosystem at methane seeps. *Mar. Ecol. Prog. Ser.* 83, 103–112.

Joblin, K.N. and Williams, A.G. (1991). Effect of cocultivation of ruminal chytrid fungi with *Methanobrevibacter smithii* on lucerne stem degradation and extracellular fungal enzyme activities. *Lett. Appl. Microbiol.* 12, 121–124.

Johnson, P.J., d'Oliveira, C.E., Gorrell, T.E. and Müller, M. (1990). Molecular analysis of the hydrogenosomal ferredoxin of the anaerobic protist *Trichomonas vaginalis*. *Proc. Natl. Acad. Sci. USA* 87, 6097–6101.

Jónasson, P.M. (1972). Ecology and production of the profundal benthos. *Oikos* Suppl. 14, 1–148.

Jones, C.W. (1985). The evolution of bacterial respiration, In: *Evolution of Prokaryotes* (ed. K.H. Schleifer and E. Stackebrandt), pp. 175–204. FEMS Symposia, No.29. Academic Press, London.

Jones, J.G. and Simon, B.M. (1985). Interaction of acetogens and methanogens in anaerobic freshwater sediments. *Appl. Environ. Microbiol.* 49, 944–948.

Jones, J.G., Gardener, S. and Simon, B.M. (1983). Bacterial reduction of ferric iron in a stratified eutrophic lake. *J. Gen. Microbiol.* 129, 131–139.

Jones, J.G., Gardener, S. and Simon, B.M. (1984). Reduction of ferric iron by heterotrophic bacteria in lake sediments. *J. Gen. Microbiol.* 130, 45–51.

Jørgensen, B.B. (1977*a*). Bacterial sulfate reduction within reduced microniches of oxidized marine sediments. *Mar. Biol.* 41, 7–17.

Jørgensen, B.B. (1977*b*). The sulfur cycle of a coastal marine sediment (Limfjorden, Denmark). *Limnol. Oceanogr.* 22, 814–832.

Jørgensen, B.B. (1978*a*). A comparison of methods for the quantification of bacterial sulfate reduction in coastal marine sediments. I. Measurement with radiotracer techniques. *Geomicrobiol. J.* 1, 11–27.

Jørgensen, B.B. (1978*b*). A comparison of methods for the quantification of sulfate reduction in coastal marine sediments. II. Calculation from mathematical models. *Geomicrobiol. J.* 1, 29–47.

Jørgensen, B.B. (1980). Seasonal oxygen depletion in the bottom waters of a Danish fjord and its effect on the benthic community. *Oikos* 34, 68–70.

Jørgensen, B.B. (1983). Processes at the sediment-water interface. In: *The Major Biogeochemical Cycles and Their Interaction.* (ed. B. Bolin and R.B. Cook), pp. 477–515. John Wiley, Chichester.

Jørgensen, B.B and Cohen, Y. (1977). Solar Lake (Sinai). 5. The sulfur cycle of the benthic cyanobacterial mats. *Limnol. Oceanogr.* 22, 657–666.

Jørgensen, B.B and Des Marais, D.J. (1986*a*). A simple fiber-optic microprobe for high resolution liquid measurements: application in marine sediments. *Limnol. Oceanogr.* 31, 1376–1383.

Jørgensen, B.B. and Des Marais, D.J. (1986*b*). Competition for sulfide among colorless and purple sulfur bacteria in cyanobacterial mats. *FEMS Microbiol. Ecol.* 38, 179–186.

Jørgensen, B.B. and Revsbech, N.P. (1989). Oxygen uptake, bacterial distribution and carbon-nitrogen-sulfur cycling in sediments from the Baltic Sea–North Sea transition. *Ophelia* 31, 29–49.

Jørgensen, B.B., Kueneu, J.G. and Cohen, Y. (1979). Microbial transformation of sulfur compounds in a stratified lake (Solar Lake, Sinai). *Limnol. Oceanogr.* 24, 799–822.

Jørgensen, B.B., Revsbech, N.P. and Cohen, Y. (1983). Photosynthesis and structure of benthic microbial mats: microelectrode and SEM studies of four cyanobacterial communities. *Limnol. Oceanogr.* 28, 1075–1093.

Kahl, A. (1928). Die Infusorien (Ciliata) der Oldesloer Salzwasserstellen. *Arch. Hydrobiol.* 19, 50–123 and 189–246.

Kahl, A. (1930–35). Urtiere oder Protozoa. I: Wimpertiere oder Ciliata (Infusoria), eine Bearbeitung der freilebenden und ectocommensalen Infusorien der Erde, unter Ausschluss der marinen Tintinnidae. *Tierwelt Dtl.* Gustav Fischer, Jena.

Kahl, A. (1931). Familie Plagiopylidae (Plagiopylina). Schew., 1896, Infusoria, Trichostomata. *Annls. Protist.* 3, 111–135.

Kandler, O. (1992). Where next with the archaebacteria? *Biochem. Soc. Symp.* 58, 195–207.

Karl, D.M. (1987). Bacterial production at deep-sea hydrothermal vents and cold seeps: evidence for chemosynthetic primary production. In: *Ecology of Microbial Communities* (ed. M. Fletcher, T.R.G. Gray and J.G.Jones), pp. 319–360. Cambridge University Press.

Kasper, H.F., Holland, A.J. and Mountfort, D.O. (1987). Simultaneous butyrate oxidation by *Syntrophomonas wolfei* and catalytic olefin reduction in absence of interspecies transfer. *Arch. Microbiol.* 147, 334–339.

Kasting, J.F. (1993). Earth's early atmosphere. *Science* 259, 920–926.

Kasting, J.F. and Donahue, T.M. (1980). The evolution of atmospheric ozone. *J. Geophys. Res.* 85, 3255–3263.

Kemp, W.M., Sampov, P.A., Garber, J., Tuttle, J. and Boynton, W.R. (1992). Seasonal depletion of oxygen from bottom waters of Chesapeake Bay: roles of benthic and planktonic respiration and physical exchange processes. *Mar. Ecol. Prog. Ser.* 85, 137–152.

Kepkay, P.E. and Nealson, K.L.T. (1987). Growth of a manganese oxidizing *Pseudomonas* sp. in continuous culture. *Arch. Microbiol.* 148, 63–67.

Kester, D.R. (1975). Dissolved gases other than CO_2. In: *Chemical Oceanography* (2nd edn) (ed. J.P. Riley and G. Skirrow), Vol. I. pp. 498–556. Academic Press, London.

Kirby, H. (1934). Some ciliates from salt marshes in California. *Arch. Protistenk.* 82, 114–133.

Kolkwitz, R. and Marsson, M. (1909). Ökologie der tierischen Saprobien. *Intern. Rev. ges. Hydrobiol. Hydrogr.* 2, 126–152.

Kondratieva, E.N., Pfennig, N. and Trüper, H.G. (1992). The phototrophic prokaryotes. In: Balows *et al.* (1992), pp. 312–330.

Krogh, A. (1941). *The Comparative Physiology of Respiratory Mechanisms.* University of Pennsylvania Press, Philadelphia.

Krumholz, L.R., Forsberg, C.W. and Veira, D.M. (1983). Association of methanogenic bacteria with rumen protozoa. *Can. J. Microbiol.* 29, 676–680.

Kudo, R.R. (1966). *Protozoology* (5th edn). Charles C. Thomas, Springfield, Illinois.

Kuenen, J.G. and Bos, P. (1989). Habitats and ecological niches of chemolitho(auto)trophic bacteria. In: Schlegel and Bowien (1989), pp. 53–80.

Kuenen, J.G., Jørgensen, B.B. and Revsbech, N.P. (1986). Oxygen microprofiles of trickling filter biofilms. *Wat. Res.* 20, 1589–1598.

Lackey, J.B. (1932). Oxygen deficiency and sewage protozoa with descriptions of some new species. *Biol. Bull.* 63, 287–295.

Lahti, C.J., d'Oliveira, C.E. and Johnson, P.J. (1992). β-succinyl-coenzyme A synthease from *Trichomonas vaginalis* is a soluble hydrogenosomal protein with an amino-terminal sequence that resembles mitochondrial presequences. *J. Bacteriol.* 174, 6822– 6830.

Lake, J.A. (1988). Origin of the eukaryotic nucleus determined by rate-invariant analysis of rRNA sequences. *Nature* 331, 184–186.

Lauterborn, R. (1901). Die 'sapropelische' Lebewelt. *Zool. Anz.* 24, 50–55.

Lee, D.L. and Atkinson, H.J. (eds). (1976). *Physiology of Nematodes* (2nd edn). Macmillan, London.

Lee, M.J. and Zinder, S.H. (1988). Isolation and characterization of a thermophilic bacterium which oxidizes acetate in syntrophic association with a methanogen and which grows acetogenically on H_2-CO_2. *Appl. Environ. Microbiol.* 54, 124–129.

Lee, M.J., Schreurs, P.J., Messer, A.C. and Zinder, S.H. (1987). Association of methanogenic bacteria with flagellated protozoa from a termite hindgut. *Current Microbiol.* 15, 337–341.

LeGall, J. and Fauque, G. (1988). Dissimilatory reduction of sulfur compounds. In: Zehnder (1988), pp. 587–639.

Leng, R.A. (1991). Improving ruminant production and reducing methane emissions from ruminants by strategic supplementation. WPA/400/1–91/004.

Leschine, S.B., Holwell, K. and Canale-Parola, E. (1988). Nitrogen fixation by anaerobic cellulolytic bacteria. *Science* 242, 1157–1159.

Leventhal, J.S. (1983). An interpretation of carbon and sulfur relationships in Black Sea sediments as indicators of environments of deposition. *Geochim. Cosmochim. Acta* 47, 133–137.

Liebmann, H. (1936). Die Ciliatenfauna der Emscherbrunnen. *Z. Hyg.* 118, 555–573.

Lindmark, D. (1980). Energy metabolism of the anaerobic protozoon *Giardia lamblia*. *Mol. Biochem. Parasitol.* 5, 291–296.

Lindmark, D.G. and Müller, M. (1973). Hydrogenosome, a cytoplasmic organelle of the anaerobic flagellate, *Tritrichomonas foetus*, and its role in pyruvate metabolism. *J. Biol. Chem.* 248, 7724–7728.

Ljungdahl, L.G. and Eriksson, K.-E. (1985). Ecology of microbial cellulose degradation. *Adv. Microb. Ecol.* 8, 237–299.

Lloyd, D., Kristensen, B. and Degn, H. (1980). The effect of inhibitors on the oxygen kinetics of terminal oxidases of *Tetrahymena pyriformis* ST. *J. Gen. Microbiol.* 121, 117–125.

Lloyd, D., Kristensen, B. and Degn, H. (1981). Oxidative detoxification of hydrogen sulphide detected by mass spectrometry in the soil amoeba *Acanthamoeba castellanii*. *J. Gen. Microbiol.* 126, 167–170.

Lloyd, D., Williams, J., Yarlett, N. and Williams, A.G. (1982). Oxygen affinities of the hydrogenosome-containing protozoa *Tritrichomonas foetus* and *Dasytricha ruminantium*, and two aerobic protozoa, determined by bacterial bioluminescence. *J. Gen. Microbiol.* 128, 1019–1022.

Lloyd, D., Hillman, K., Yarlett, N. and Williams, A.G. (1989). Hydrogen production by rumen holotrich protozoa: effects of oxygen and implications for metabolic control by in situ conditions. *J. Protozool.* 36, 205–213.

Lovely, D.R. and Klug, M.J. (1982). Intermediary metabolism of organic matter in the sediments of a eutrophic lake. *Appl. Environ. Microbiol.* 43, 552–560.

Lovely, D.R. and Klug, M.J. (1983). Methanogenesis from methanol and methylamines and acetogenesis from hydrogen and carbon dioxide in the sediments of a eutrophic lake. *Appl. Environ. Microbiol.* 45, 1310–1315.

Lupton, F.S., Conrad, R. and Zeikus, J.G. (1984). Physiological function of hydrogen metabolism during growth of sulfidogenic bacteria on organic substrates. *J. Bacteriol.* 159, 843–849.

Madigan, M.T. (1988). Microbiology, physiology, and ecology of phototrophic bacteria. In: Zehnder(1988), pp. 39–111.

Mah, R.A. (1982). Methanogenesis and methanogenic partnerships. *Phil. Trans. Roy. Soc. (London) B* 297, 599–616.

Maier, S. and Gallardo, V.A. (1984). Nutritional characteristics of two marine thioplocas determined by autoradiography. *Arch. Microbiol.* 139, 218–220.

Margulis, L. (ed.) (1970). *Origin of Eucaryotic Cells.* Yale University Press, New Haven.

Margulis L. (ed.) (1993). *Symbiosis in Cell Evolution.* (2nd edn). W.H. Freeman, San Francisco.

Martens, C.S. and Berner, R.A. (1974). Methane production in the interstitial waters of sulfate-depleted marine sediments. *Science* 185, 1167–1169.

Marvin-Sikkema, F.D., Gomes, T.M.P., Grivet, J.-P., Gottschal, J.C. and Prins, R.A. (1993). Characterization of hydrogenosomes and their role in glucose metabolism of *Neocallimastix* sp. L2. *Arch. Microbiol.* 160, 388–396.

McInerney, M.J. (1986). Transient and persistent associations among prokaryotes. In: *Bacteria in Nature* (ed. J.S. Poindexter and E.R. Leadbetter), Vol. II, pp. 293–338. Plenum Press, New York.

McInerney, M.J. (1992). The genus *Syntrophomonas* and other syntrophic anaerobes. In: Balows et al. (1992), pp. 2048–2057.

McInerney, M.J., Bryant, M.P., Hespell, R.B. and Costerton, J.W. (1981). *Syntrophomonas wolfei* gen. nov. sp. nov., an anaerobic, syntrophic, fatty-acid oxidizing bacterium. *Appl. Environ. Microbiol.* 41, 1029–1039.

McMahon, B.R. (1988). Physiological responses to oxygen depletion in intertidal animals. *Amer. Zool.* 28, 39–53.

Messer, A.C. and Lee, M.J. (1989). Effect of chemical treatments on methane emission by the hindgut microbiota in the termite *Zootermopsis angusticollis*. *Microbiol. Ecol.* 18, 275–284.

Meyers, H.B., Fossing, H. and Powell, E.N. (1987). Microdistribution of interstitial meiofauna, oxygen and sulfide gradients, and the tubes of macro-infauna. *Mar. Ecol. Prog. Ser.* 35, 223–241.

Meyers, M.B., Powell, E.N. and Fossing, H. (1988). Movement of oxybiotic and thiobiotic meiofauna in response to changes in pore-water oxygen and sulfide gradients around macro-infaunal tubes. *Mar. Biol.* 98, 395–414.

Miller, S.L. (1953). A production of amino acids under possible primitive Earth conditions. *Science* 117, 528–529.

Miller, S.L. and Orgel, L.E. (ed.) (1974). *The Origins of Life on the Earth.* Prentice-Hall, Englewood Cliffs, New Jersey.

Millero, F.J. (1986). The thermodynamics and kinetics of the hydrogen sulfide system in natural waters. *Mar. Chem.* 18, 121–147.

Miracle, M.R., Vincente, E. and Pedrós-Alió, C. (1992). Biological studies of Spanish meromictic and stratified karstic lakes. *Limnetica* 8, 59–77.

Moir, R.J. (1965). The comparative physiology of ruminant-like animals. In: *Physiology of Digestion in the Ruminant* (ed. R.W. Dougherty), pp.1–23. Butterworths, Washington.

Möller, D., Schauder, R., Fuchs, G. and Thauer, R.K. (1987). Acetate oxidation to CO_2 via a citric acid cycle involving an ATP-citrate lyase: a mechanism for the synthesis of ATP via substrate level phosphorylation in *Desulfobacter postgatei* growing on acetate and sulfate. *Arch. Microbiol.* 148, 202–207.

Mortimer, C.H. (1941–42). The exchange of dissolved substances between mud and water in lakes. I-II. *J. Ecol.* 29, 280–329 and 30, 147–201.

Moss, A. (1992). Methane from ruminants in relation to global warming. *Chem. Ind.* 9, 334–336.

Müller, M. (1980). The hydrogenosome. In: *The Eukaryotic Microbial Cell* (ed. G.W. Gooday, D. Lloyd, and A.P.J. Trinci), pp. 127–142. Cambridge University Press.

Müller, M. (1988). Energy metabolism of protozoa without mitochondria. *Ann. Rev. Microbiol.* 42, 465–488.

Müller, M. (1992). Energy metabolism of ancestral eukaryotes: a hypothesis based on the biochemistry of amitochondriate parasitic protists. *BioSystems* 28, 33–40.

Müller, M. (1993). The hydrogenosome. *J. Gen. Microbiol.* 139, 2879–2889.

Munn, E.A., Orpin, C.G. and Greenwood, A. (1988). The ultrastructure and possible relationships of four obligate anaerobic chytridiomycete fungi from the rumen of sheep. *BioSystems* 22, 67–81.

Murray, R.M., Marsh, H., Heinsok, G.E. and Spain, A.V. (1977). The role of the midgut caecum and large intestine in the digestion of sea grasses by the dugong (Mammalia: *Sirenia*). *Comp. Biochem. Physiol.* A 56, 7–10.

Mylnikov, A.P. (1991). Diversity of flagellates without mitochondria. In: Patterson and Larsen (1991), pp. 149–158.

Nedwell, P.B. and Gray, T.R.G. (1987). Soils and sediments as materials for microbial growth. In: *Ecology of Microbial Communities* (ed. M. Fletcher, T.R.G. Gray and J.G. Jones), pp. 21–54. Cambridge University Press.

Nelson,D.C. (1989). Physiology and biochemistry of filamentous sulfurbacteria. In: Schlegel and Bowien (1989), pp. 219–238.

Nelson, D.C., Revsbech, N.P. and Jørgensen, B.B. (1986a). The microoxic/anoxic niche of *Beggiatoa* spp.: microelectrode survey of marine and freshwater strains. *Appl. Environ. Microbiol.* 52, 161–168.

Nelson, D.C., Jørgensen, B.B. and Revsbeck, N.P. (1986b). Growth pattern and yield of a chemoautotrophic *Beggiatoa* sp. in oxygen-sulfide microgradients. *Appl. Environ. Microbiol.* 52, 225–233.

Nethe-Jaenchen, R. and Thauer, R.K. (1984). Growth yields and saturation constant of *Desulfovibrio vulgaris* in chemostat culture. *Arch. Microbiol.* 137, 236–240.

Nicholas, W.L. (ed.) (1975). *The Biology of Free-living Nematodes.* Clarendon Press, Oxford.

Nicholas, W.L. (1991). Interstitial meiofauna. In: Bryant (1991), pp. 129–145.

Nicholson, J.A.M., Stolz, J.F. and Pierson, B.K. (1987). Structure of a microbial mat at Great Sippewissett Marsh, Cape Cod, Massachusetts. *FEMS Microbiol. Ecol.* 45, 343–364.

Nursall, J.R. (1959). Oxygen as a prerequisite to the origin of the Metazoa. *Nature* 183, 1170–1172.

Nuss, B. (1984). Ultrastrukturelle und ökophysiologische Untersuchungen auf kristalloiden Einschlüssen der Muskeln eines sulfidtoleranten limnischen Nematoden (*Tobrilus gracilis*). *Veröff. Inst. Meeresforsch. Bremerh.* 20, 3–15.

Odelson, D.A. and Breznak, J.A. (1983). Volatile fatty acid production by the hindgut microbiota of xylophagus termites. *Appl. Environ. Microbiol.* 45, 1602–1613.

Odelson, D.A. and Breznak, J.A. (1985). Nutrition and growth characteristics of *Trichomitopsis termopsidis*, a cellulolytic protozoan from termites. *Appl. Env. Microbiol.* 49, 614–621.

Odom, J.M. and Peck, H.D. (1981). Hydrogen cycling as a general mechanism for energy coupling in the sulfate reducing bacteria, *Desulfovibrio* sp. *FEMS Microbiol. Lett.* 12, 47–50.

Oeschger, R. and Schmaljohann, R. (1988). Association of various types of epibacteria with *Halicryptus spinulosus* (Priapulida). *Mar. Ecol. Prog. Ser.* 48, 285–293.

Oeschger, R. and Vetter, R.D. (1992). Sulfide detoxification and tolerance in *Halicryptus spinulosus* (Priapulida): a multiple strategy. *Mar. Ecol. Prog. Ser.* 86, 167–179.

Ogimoto, K. and Imai, S. (ed.) (1981). *Atlas of Rumen Microbiology.* Japan Scientific Societies Press, Tokyo. 231pp.

Olson, J.M. and Pierson, B.K. (1987). Evolution of reaction centers in photosynthetic centers. *Origins of Life* 17, 419–430.

Oparin, A.I. (ed.) (1953). *The Origin of Life.* (2nd ed.). Dover Publications, New York.

Oremland, R.S. (1988). Biochemistry of methanogenic bacteria. In: Zehnder (1988), pp. 641–705.

Oremland, R.S and Capone, D.G. (1987). Use of the 'specific' inhibitors in biogeochemistry and microbial ecology. *Adv. Microbiol. Ecol.* 10, 285–383.

Oremland, R.S. and Polcin, S. (1982). Methanogenesis and sulfate reduction: competitive and noncompetitive substrates in estuarine sediments. *Appl. Environ. Microbiol.* 44, 1270–1276.

Oremland, R.S. and Taylor, B.F. (1978). Sulfate reduction and methanogenesis in marine sediments. *Geochim. Cosmochim. Acta* 42, 209–214.

Oren, A. and Blackburn,T.H. (1979). Estimation of sediment denitrification rates at *in situ* nitrate concentration. *Appl. Environ. Microbiol.* 37, 174–176.

Orgel, L.E. (1986). RNA catalysis and the origins of life. *J. Theoret. Biol.* 123, 127–149.

Ott, J.A. and Schiemer, F. (1973). Respiration and anaerobiosis of free-living nematodes from marine and limnic sediments. *Neth. J. Sea Res.* 7, 233–243.

Ott, J.A., Rieger, G., Rieger, R. and Enderes, F. (1982). New mouthless interstitial worms from the sulfide system: symbiosis with prokaryotes. *PSZNI-Mar. Ecol.* 38, 313–333.

Paget, T.A. and Lloyd, D. (1990). *Trichomonas vaginalis* requires traces of oxygen and high concentrations of carbon dioxide for optimal growth. *Mol. Biochem. Parasitol.* 41, 65–72.

Paget, T.A., Jarroll, E.L., Manning, P., Lindmark, D.G. and Lloyd, D. (1989). Respiration in cysts and trophozoites of *Giardia muris*. *J. Gen. Microbiol.* 135, 145–154.

Paget, T.A., Kelly, M.L., Jarroll, E.L., Lindmark, D.G. and Lloyd, D. (1993). The effects of oxygen on fermentation in *Giardia lamblia*. *Mol. Biochem. Parasitol.* 57, 65–72.

Pankhania, I.P., Gow, L.A. and Hamilton, W.A. (1986). The effect of hydrogen on the growth of *Desulfovibrio vulgaris* (Hildenborough) on lactate. *J. Gen. Microbiol.* 132, 3349–3356.

Patterson, D.J. and Larsen, J. (ed.) (1991). *The Biology of Free-living Flagellates.* Oxford University Press.

Pearsall, W.H. and Mortimer, C.H. (1939). Oxidation-reduction potentials in water-logged soils, natural waters and muds. *J. Ecol.* 27, 483–501.

Pearson, T.H. and Rosenberg, R. (1978). Macrobenthic succession in relation to organic enrichment and pollution of the marine environment. *Oceanogr. Mar. Biol. Ann. Rev.* 16, 229–311.

Peck, H.D., LeGall, J., Lespinat, P.A., Berlier, Y. and Fauque, G. (1987). A direct demonstration of hydrogen cycling by *Desulfovibrio vulgaris* employing membrane-inlet mass spectrometry. *FEMS Microbiol. Lett.* 40, 295–299.

Pfennig, N. (1980). Syntrophic mixed cultures and symbiotic consortia with phototrophic bacteria: a review. In: *Anaerobes and Anaerobic Infections.* (ed. G. Gottschalk, N. Pfennig and H. Werner), pp. 127–131. Gustav Fischer-Verlag, Stuttgart.

Pfennig, N. (1989). Ecology of phototrophic purple and green sulfur bacteria. In: Schlegel and Bowien (1989), pp. 97–116.

Phelps, T.J. and Zeikus, J.G. (1984). Influence of pH on terminal carbon metabolism in anoxic sediments from a mildly acidic lake. *Appl. Environ. Microbiol.* 48, 1088–1095.

Pierson, B.K. and Olson, J.M. (1989). Evolution of photosynthesis in anoxygenic photosynthetic prokaryotes. In: Cohen and Rosenberg (1989), pp. 402–427.

Pierson, B.K., Oesterle, A. and Murphy, G.L. (1987). Pigments, light penetration, and photosynthetic activity in the multilayered microbial mats of Great Sippewisset Salt Marsh, Massachusetts. *FEMS Microbiol. Ecol.* 45, 365–376.

Pirt, S.J. (1966). The maintenance energy of bacteria in growing cultures. *Proc. Roy. Soc. (London) B* 163, 224–231.

Por, F.D. and Masry, D. (1968). Survival of a nematode and an oligochaete species in the anaerobic benthal Lake Tiberias. *Oikos* 19, 388–391.

Postgate, J.R. (ed.) (1982). *The Fundamentals of Nitrogen Fixation.* Cambridge University Press.

Powell, G.E. (1984). Equalization of specific growth rates for syntrophic associations in batch culture. *J. Chem. Technol. Biotechnol.* 34B, 97–100.

Powell, E. (1989). Oxygen sulfide and diffusion: why thiobiotic meiofauna must be sulfide-insensitive first-order respires. *J. Mar. Res.* 47, 887–932.

Powell, M.A. and Somero, G.N. (1986). Hydrogen sulfide oxidation is coupled to oxidative phosphorylation in mitochondria of *Solemya reidi*. *Science* 233, 563–566.

Powell, E.N., Crenshaw, M.A. and Rieger, R.M. (1979). Adaptations to sulfide in the meiofauna of sulfide systems. I. ^{35}S-sulfide accumulations and the presence of a sulfide detoxification system. *J. Exp. Mar. Ecol.* 37, 57–76.

Powell, E.N., Crenshaw, M.A. and Rieger, R.M. (1980). Adaptations to sulfide in sulfide-system meiofauna. Endproducts of sulfide detoxification in three turbellarians and a gastrotrich. *Mar. Ecol. Prog. Ser.* 2, 169–177.

Powell, E.N., Bright, T.J., Woods, A. and Gittings, S. (1983). Meiofauna and the thiobios in the East Flower Garden brine seep. *Mar. Biol.* 73, 264–283.

Psenner, R. and Schlott-Idl, K. (1985). Trophic relationships between bacteria and protozoa in the hypolimnion of a meromictic mesotrophic lake. *Hydrobiologia* 121, 111–120.

Purcell, E.M. (1977). Life at low Reynolds number. *Am. J. Physics* 45, 3–11.

Rahat, M., Judd, J. and van Eys J. (1964). Oxidation of reduced diphosphopyridine nucleotide by *Tetrahymena pyriformis* preparations. *J. Biol. Chem.* 239, 3537–3545.

Raikov, I.B. (1971). Bactéries épizoiques et mode de nutrition du ciliés psammophile *Kentrophoros fistolosum* Fauré-Fremiet (étude du microscope électronique). *Protistologica* 7, 365–378.

Raikov, I.B. (1974). Étude ultrastructurale de bactéries épizoiques et endozoiques de *Kentrophoros latum* Raikov, ciliés holotriche mésopsammique. *Cah. Biol. Mar.* 15, 379–393.

Reeburgh, W.S. (1980). Anaerobic methane oxidation: rate depth distribution in Skan Bay sediments. *Earth Planet. Sci. Lett.* 47, 345–352.

Reeves, R.E. (1984). Metabolism of *Entamoeba histolytica* Schaudinn, 1903. *Adv. Parasitol.* 23, 105–142.

Reise, K. (1981a). High abundance of small zoobenthos around biogenic structures in tidal sediments of the Waddensea. *Helgoländer wiss. Meeresunters.* 34, 413–425.

Reise, K. (1981b). Gnathostomulida abundant alongside polychaete burrows. *Mar. Ecol. Prog. Ser.* 6, 329–333.

Reise, K. (1984). Free-living Platyhelminthes (Turbellaria) of a marine sand flat: an ecological study. *Microfauna Marina* 1, 1–62.

Reise, K. and Ax, P. (1979). A meiofaunal 'thiobios' limited to the anaerobic sulfide system of marine sand does not exist. *Mar. Biol.* 54, 225–237.

Revsbech, N.P. and Jørgensen, B.B. (1986). Microelectrodes: their use in microbial ecology. *Adv. Microbiol. Ecol.* 9, 293–352.

Revsbech, N.P., Jørgensen, B.B. and Blackburn, T.H. (1980). Oxygen in the sea bottom measured with a microelectrode. *Science* 207, 1355–1356.

Revsbech, N.P., Jørgensen, B.B., Blackburn, T.H. and Cohen, Y. (1983). Microelectrode studies of the photosynthesis and O_2, H_2S and pH profiles of a microbial mat. *Limnol. Oceanogr.* 28, 1062–1074.

Revsbech, N.P., Sørensen, J., Blackburn, T.H. and Lomholt,J.P. (1980). Distribution of oxygen in marine sediments measured with microelectrodes. *Limnol. Oceanogr.* 25, 403–411.

Rheinheimer, G., Gocke, K. and Hoppe, H.-G. (1989). Vertical distribution of microbiological and hydrographic-chemical parameters in different areas of the Baltic Sea. *Mar. Ecol. Prog. Ser.* 52, 55–70.

Rhoads, D.C. and Morse, J.W. (1971). Evolutionary and ecological significance of oxygen-deficient marine basins. *Lethaia* 4, 413–428.

Richards, F.A. (1975). The Cariaco Basin (Trench). *Oceanogr. Mar. Biol. Ann. Rev.* 13, 11–67.

Ridder, C. De, Jangoux, M. and Vos, L. De (1985). Description and significance of a peculiar intradigestive symbiosis between bacteria and a deposit-feeding echinoid. *J. Exp. Mar. Biol. Ecol.* 91, 65–76.

Robertson, L.A. and Kuenen, J.G. (1983). *Thiosphaera pantotropha* gen. nov. sp. nov., a facultatively anaerobic, facultatively autotrophic sulphur bacterium. *J. Gen. Microbiol.* 129, 2847–2855.

Robertson, L.A. and Kuenen, J.G. (1984). Aerobic denitrification: a controversy revived. *Arch. Microbiol.* 139, 351–354.

Roden, E.E. and Lovely, D.R. (1993). Dissimilatory Fe(III) reduction by the marine microorganism *Desulfuromonas acetoxidans*. *Appl. Environ. Microbiol.* 59, 734–742.

Römmermann, D. and Friedrich, B. (1985). Denitrification by *Alcaligenes eutrophus* is plasmid dependent. *J. Bacteriol.* 162, 852–854.

Runnegar, B. (1991). Oxygen and the early evolution of the metazoa. In: Bryant (1991), pp. 65–87.

Russell, J. and Baldwin, R.L. (1979*a*). Comparison of substrate affinities among several rumen bacteria: a possible determinant of rumen bacterial competition. *Appl. Environ. Microbiol.* 37, 531–536.

Russell, J. and Baldwin, R.L. (1979*b*). Comparison of maintenance energy expenditures and growth yields among several rumen bacteria grown in continuous culture. *Appl. Environ. Microbiol.* 37, 537–543.

Russell, J.R., Delfino, F.J. and Baldwin, R.L. (1979). Effects of combinations of substrates on maximum growth rates of several rumen bacteria. *Appl. Environ. Microbiol.* 37, 544–549.

Ryley, J.F. (1952). Studies on the metabolism of the protozoa. 3. Metabolism of the ciliate *Tetrahymena pyriformis* (*Glaucoma piriformis*). *Biochem. J.* 52, 483–492.

Schauder, R. and Kröger, A. (1993). Bacterial sulphur respiration. *Arch. Microbiol.* 159, 491–497.

Schenk, H.E.A., Bayer, M.G. and Zook, D. (1987). Cyanelles from symbiont to organelle. *Ann. N.Y. Acad. Sci.* 503, 151–167.

Schiemer, F. (1973). Respiration rates of two species of Gnathostomulids. *Oecologia* (Berl.) 13, 403–406.

Schiemer, F., Novak, R. and Ott, J.A. (1990). Metabolic studies on thiobiotic free-living nematodes and their symbiotic microorganisms. *Mar. Biol.* 106, 129–137.

Schink, B. (1988). Principles and limits of anaerobic degradation: environmental and technological aspects. In: Zehnder (1988), pp. 771–846.

Schink, B. (1992*a*). Syntrophism among prokaryotes. In: Balows et al. (1992), pp. 276–299.

Schink, B. (1992*b*). The genus *Propionigenium*. In: Balows *et al.* (1992), pp. 3948–3951.

Schink, B. and Zeikus, J.G. (1980). Microbial methanol formation: a major end-product of pectin metabolism. *Curr. Microbiol.* 4, 387–389.

Schlegel, H.G. (1986). *General Microbiology* (6th edn). Cambridge University Press.

Schlegel, H.G. (1989). Aerobic hydrogen-oxidizing (knallgas) bacteria. In: Schlegel and Bowien (1989), pp. 305–329.

Schlegel, H.G. and Bowien, B. (eds). (1989). *Autotrophic bacteria.* Science Technical Publishers Madison Springer-Verlag, Berlin.

Schlegel, H.G. and Jannasch, H.W. (1992). Prokaryotes and their habitats. In: Balows *et al.* (1992), pp. 75–125.

Schlegel, M. (1991). Protist evolution and phylogeny as descerned from small subunit ribosomal RNA sequence comparisons. *Europ. J. Protistol.* 27, 207–219.

Schmaljohann, R. and Flügel, H. (1987). Methane-oxidizing bacteria in Pogonophora. *Sarsia* 72, 91–98.

Schneider, B., Nies, A. and Friedrich, B. (1988). Transfer and expression of lithoautotrophy and denitrification in a host lacking these metabolic activities. *Appl. Environ. Microbiol.* 54, 3173–3176.

Schopf, J.W. (ed.) (1983). *Origin and Evolution of Earth's Earliest Biosphere: Its Origin and Evolution.* Princeton University Press, Princeton, New Jersey.

Schopf, J.W. (1992). Palaeobiology of the Archean. In: Schopf and Klein (1992), pp. 25–40.

Schopf, J.W. and Klein, C. (ed.) (1992). *The Proterozoic Biosphere.* Cambridge University Press.

Schopf, J.W. and Walter, M.R. (1983). Archaean microfossils: new evidence of ancient microbes. In: Schopf (1983), pp. 214–239.

Schöttler, U. (1977). NADH-generating reactions in anaerobic *Tubifex* mitochondria. *Comp. Biochem. Physiol.* 58 B, 261–265.

Schöttler, U. and Bennet, E.M. (1991). Annelids. In: Bryant (1991), pp. 165–185.

Schrago, E. and Elson, C. (1980). Intermediary metabolism of *Tetrahymena.* In: *Biochemistry and Physiology of Protozoa* (2nd edn) (ed. M. Levandowsky and S.H. Hutner), Vol. 3, pp. 287–312. Academic Press. New York.

Schroff, G. and Zebe, E. (1980). The anaerobic formation of propionic acid in the mitochondria of the lugworm *Arenicola marina. J. Comp. Physiol.* 138, 35–41.

Schuster, P. (1981). Prebiotic evolution. In: *Biochemical evolution* (ed. H. Gutfreund), pp. 15–87. Cambridge University Press.

Scranton, M.I. and Brewer, P.G. (1977). Occurrence of methane in the near-surface waters of the western subtropical North-Atlantic. *Deep-Sea Res.* 24, 127–138.

Searcy, D.G. (1987). Phylogenetic and phenotypic relationships between the eukaryotic nucleocytoplasm and thermophilic archaebacteria. *Ann. N.Y. Acad. Sci.* 503, 168–179.

Searcy, D.G. and Hixon, W.G. (1991). Cytoskeletal origins in sulfur-metabolizing archaebacteria. *BioSystems.* 25, 1–11.

Senior, E., Lindström, E.B., Banat, I.M. and Nedvelle, D.B. (1982). Sulfate reduction and methanogenesis in the sediment of a saltmarsh on the east coast of the United Kingdom. *Appl. Environ. Microbiol.* 43, 987–996.

Shanks, A.L. and Reeder, M.L. (1993). Reducing microzones and sulfide production in marine snow. *Mar. Ecol. Prog. Ser.* 96, 43–47.

Sillén, L.G. (1966). Regulation of O_2, N_2 and CO_2 in the atmosphere: thoughts of a laboratory chemist. *Tellus* 18, 198–206.

Small, E.B. and Lynn, D.H. (1981). A new macrosystem for the phylum Ciliophora Doflein, 1901. *BioSystems* 14, 387–401.

Sogin, M.L. (1991). Early evolution and the origin of eukaryotes. *Curr. Opinion. Genet. Dev.* 1, 457–463.

Sogin, M.L., Elmwood, H.S. and Gundersen, J.H. (1986). Evolutionary diversity of eukaryotic small subunit rRNA genes. *Proc. Natl. Acad. Sci. USA* 83, 1383–1387.

Sørensen, J. (1982). Reduction of ferric iron in anaerobic, marine sediment and interaction with reduction of nitrate and sulfate. *Appl. Environ. Microbiol.* 43, 319–324.

Sørensen, J., Jørgensen, B.B. and Revsbech, N.P. (1979). A comparison of oxygen, nitrate and sulfate respiration in coastal marine sediments. *Microbiol. Ecol.* 5, 105–115.

Sørensen, J., Christensen, D. and Jørgensen, B.B. (1981). Volatile fatty acids and hydrogen as substrates for sulfate-reducing bacteria in anaerobic marine sediment. *Appl. Environ. Microbiol.* 42, 5–11.

Sorokin, Y.I. (1965). On the trophic role of chemosynthesis and bacterial biosynthesis in water bodies. *Mem. Ist. ital. Idrobiol.* 18 suppl, 187–205.

Sorokin, Y.I. (1970). Interrelations between sulphur and carbon turnover in meromictic lakes. *Arch. Hydrobiol.* 66, 391–446.

Sorokin, Y.I. (1972). The bacterial population and the process of hydrogen sulphide oxidation in the Black Sea. *J. Cons., Int. Explor. Mer.* 34, 423–455.

Southward, E.C. (1982). Bacterial Symbionts in Pogonophora. *J. Mar. Biol. Assoc. UK.* 62, 889–906.

Southward, E.C. (1987). Contribution of symbiotic chemoautotrophs to the nutrition of benthic invertebrates. In: *Microbes in the Sea* (ed. M.A. Sleigh), pp. 83–118. John Wiley, New York.

Stackebrandt, E. (1992). Unifying phylogeny and phenotypic diversity. In: Balows *et al.* (1992), pp. 19–47.

Stanier, R.Y., Ingraham, J.L., Wheelis, M.L. and Painter, P.R. (1987). *General Microbiology*, (5th edn) Macmillan, London.

Steinbüchel, A. and Müller, M. (1986a). Glycerol, a metabolic end product of *Trichomonas vaginalis* and *Tritrichomonas foetus*. *Mol. Biochem. Parasit.* 20, 45–55.

Steinbüchel, A. and Müller, M. (1986b). Anaerobic pyruvate metabolism of *Tritrichomonas foetus* and *Trichomonas vaginalis* hydrogenosomes. *Mol. Biochem. Parasit.* 20, 57–65.

Stewart, K.D. and Mattox, K.R. (1984). The case for a polyphyletic origin of mitochondria: morphological and molecular comparisons. *J. Mol. Evol.* 21, 54–57.

Stoecker, D.K., Michaels, A.E. and Davis, L.H. (1987). Large proportion of marine planktonic ciliates found to contain functional chloroplasts. *Nature*, 326, 790–792.

Storey, K.B. and Storey, J.M. (1990). Metabolic rate depression and biochemical adaptation in anaerobiosis, hibernation and estivation. *Quart. Rev. Biol.* 65, 145–174.

Stouthamer, A.H. (1988). Dissimilatory reduction of oxidised nitrogen compounds. In: Zehnder (1988), pp. 245–303.

Stouthamer, A.H., van't Riet, J. and Oltman, L.F. (1980). Respiration with nitrate as acceptor. In: *Diversity of Bacterial Respiratory Systems* (ed. C.J. Knowles), pp. 19–48. CRC Press, Boca Raton.

Strohhäcker, J., de Graaf, A.A., Schoberth, S.M., Wittig, R.M. and Sahm, H. (1993). [31]P nuclear magnetic resonance studies of ethanol inhibition in *Zymomonas mobilis*. *Arch. Microbiol.* 159, 484–490.

Stryer, L. (ed.) (1988). *Biochemistry.* (3rd edn). W.H. Freeman, New York.

Stumm, W. and Morgan, J.J (ed.) (1970). *Aquatic Chemistry.* John Wiley, New York.

Sturr, M.G. and Marquis, R.E. (1992). Comparative acid tolerances and inhibitor sensitivities of isolated F-ATPases of oral lactic acid bacteria. *Appl. Environ. Microbiol.* 58, 2287–2291.

Swedmark, B. (1964). The interstitial fauna of marine sand. *Biol. Rev.* 39, 1–41.

Tatton, M.J., Archer, D.B., Powell, G.E. and Parker, M.L. (1989). Methanogenesis from ethanol by defined mixed continuous cultures. *Appl. Environ. Microbiol.* 55, 440–445.

Taylor, A.C. (1990). Respiratory responses to environmental conditions in intertidal prawns. In: *Animal Nutrition and Transport Processes* (eds. J.-P. Truchot and B. Lahlou), pp. 104–118. Karger, Basel.

Thauer, R.K., Jungerman, K. and Decker, K. (1977). Energy conservation in chemotrophic bacteria. *Bact. Rev.* 41, 100–180.

Thiele, J.H., Chartrain, M. and Zeikus, J.G. (1988). Control of interspecies electron flow during anaerobic digestion: role of floc formation in syntrophic methanogenesis. *Appl. Environ. Microbiol.* 54, 10–19.

Tiedje, J.M. (1988). Ecology of denitrification and dissimilatory nitrate reduction to ammonium. In: Zehnder (1988), pp. 179 — 244.

Trüper, H.G. and Fischer, U. (1982). Anaerobic oxidation of sulphur compounds as electron donors for bacterial photosynthesis. *Phil. Trans Roy. Soc. (London) B* 298, 529–542.

Trüper, H.G. and Genovese, S. (1968). Characterization of photosynthetic sulfur bacteria causing red water in Lake Faro (Messina, Sicily). *Limnol. Oceanogr.* 13, 225–232.

Tyson, R.V. and Pearson, T.H. (ed.) (1991). *Modern and Ancient Continental Shelf Anoxia.* The Geological Society, London.

Ultsch, G.R. and Jackson, D.C. (1982). Long-term submergence at 3°C of the turtle, *Chrysemys picta bellii*, in normoxic and severely hypoxic water. *J. Exp. Biol.* 96, 11–28.

Ultsch, G.R., Ott, M.E. and Heisler, N. (1980). Standard metabolic rate, critical oxygen tension and aerobic scope for spontaneous activity of trout (*Salmo gairdneri*) and carp (*Cyprinus carpio*) in acidified water. *Comp. Biochem. Physiol.* 67 A, 324–335.

Ultsch, G.R., Jackson, D.C. and Moalli, R. (1981). Metabolic oxygen conformity among lower vertebrates: the toadfish revisited. *J. Physiol.* 142, 439–443.

Vanderborght, J.P. and Billen, G. (1975). Vertical distribution of nitrate concentration in interstitial water of marine sediments with nitrification and denitrification. *Limnol. Oceanogr.* 20, 953–961.

Van Bruggen, J.J.A., Stumm, C.K. and Vogels, G.D. (1983). Symbiosis of methanogenic bacteria and sapropelic protozoa. *Arch. Microbiol.* 136, 89–96.

Van Bruggen, J.J.A., Zwart, K.B., van Assema, R.M., Stumm, C.K. and Vogels, G.D. (1984). *Methanobacterium formicicum*, an endosymbiont of the anaerobic ciliate *Metopus striatus* McMurrich. *Arch. Microbiol.* 139, 1–7.

Van Bruggen, J.J.A., Stumm, C.K., Zwart, K.B. and Vogels, G.D. (1985). Endosymbiotic methanogenic bacteria of the sapropelic amoeba *Mastigella*. *FEMS Microbiol. Ecol.* 31, 187–192.

Van Bruggen, J.J.A., Zwart, K.B., Herman, J.G.F., Van Hove, E.M., Assema, R.M., Stumm, C.K. and Vogels, G.D. (1986). Isolation and characterisation of *Methanoplanus endosymbiosus* sp. nov. an endosymbiont of the marine sapropelic ciliate *Metopus contortus* Quennerstedt. *Arch. Microbiol.* 144, 367–374.

Van Bruggen, J.J.A., Van Rens, G.L.M., Geertman, E.J.M., Zwart, K.B., Stumm, C.K. and Vogels, G.D. (1988). Isolation of a methanogenic endosymbiont of the sapropelic amoeba *Pelomyxa palustris* Greef. *J. Protozool.* 35, 20–23.

Van Gemerden, H. (1974). Coexistence of organisms competing for the same substrate: an example among purple sulfur bacteria. *Microbiol. Ecol.* 1, 104–119.

Van Gemerden, H. (1993). Microbial mats: a joint venture. *Mar. Geol.* 113, 3–25.

Van Gemerden, H. and Beeftink, H.H. (1983). Coexistence of *Chlorobium* and *Chromatium* in a sulfide-limited continous culture. *Arch. Microbiol.* 129, 32–34.

Van Gemerden, H. and de Wit, R. (1989). Phototrophic and chemotrophic growth of the purple sulfur bacterium *Thiocapsa roseopersicina*. In: Cohen and Rosenberg (1989), pp. 313–319.

Veldkamp, H. and Jannasch, H.W. (1972). Mixed culture studies with the chemostat. *J. Appl. Chem. Biotechnol.* 22, 105–123.

Vetter, R.D., Powell, M.A. and Somero, G.N. (1991). Metazoan adaptations to hydrogen sulphide. In: Bryant (1991). pp. 109–128.

Vicente, E., Rodrigo, M.A., Camacho, A. and Miracle, M.R. (1991). Phototrophic prokaryotes in a karstic sulphate lake. *Verh. Intern. Verein. Limnol.* 24, 998–1004.

Vismann, B. (1990). Sulfide detoxification and tolerance in *Nereis (Hediste) diversicolor* and *Nereis (Neanthes) virens* (Annelida: Polychaeta). *Mar. Ecol. Prog. Ser.* 59, 229–238.

Vismann, B. (1991a). Sulfide tolerance: physiological mechanisms and ecological implications. *Ophelia* 34, 1–27.

Vismann, B. (1991b). The physiology of sulfide detoxification in the isopod *Saduria (Mesidotea) entomon* (L.). *Mar. Ecol. Prog. Ser.* 76, 283–293.

Visscher, P.T., Prins, R.A. and van Gemerden, H. (1992a). Rates of sulfate reduction and thiosulfate consumption in a marine microbial mat. *FEMS Microbiol. Ecol.* 86, 283–294.

Visscher, P.T., van den Ende, F.P., Schaub, B.E.M. and van Gemerdem, H. (1992b). Competition between anoxygenic phototrophic bacteria and colorless sulfur bacteria in a microbial mat. *FEMS Microbiol Ecol.* 101, 57–58.

Vogels, G.D., Hoppe, W.F. and Stumm, C.K. (1980) Association of methanogenic bacteria with rumen ciliates. *Appl. Environ. Microbiol.* 40, 608–612.

Vogels, G.D., Keltjens, J.T. and van der Drift, C. (1988). Biochemistry of methane production. In: Zehnder (1988), pp. 707–770.

Wagener, S. and Pfennig, N. (1987). Monoxenic culture of the anaerobic ciliate *Trimyema compressum* Lackey. *Arch. Microbiol.* 149, 9–11.

Wagener, S., Schulz, S. and Hanselmann, K. (1990). Abundance and distribution of anaerobic protozoa and their contribution to methane production in Lake Cadagno (Switzerland). *FEMS Microbiol. Lett.* 74, 39–48.

Wald, G. (1966). On the nature of cellular respiration. In: *Current Aspects of Biochemical Energetics* (ed. N.O. Kaplan and E.P. Kennedy), pp. 27–32. Academic Press, New York.

Walker, J.C.G. (1980). The influence of life on evolution of the atmosphere. *Life Sci. Space Res.* 8, 89–100.

Walter, M.R. (1983). Archean stromatolites: evidence of the Earth's earliest benthos. In: Schopf (1983), pp. 187–213.

Ward, D.M. and Winfrey, M.R. (1985). Interactions between methanogenic and sulfate-reducing bacteria in sediments. *Adv. Aquatic Microbiol.* 3, 141–179.

Ward, D.M., Weller, R., Shiea, J., Castenholz, R.W. and Cohen, Y. (1989). Hot spring microbial mats: anoxygenic and oxygenic mats of possible evolutionary significance. In: Cohen and Rosenberg (1989), pp. 3–15.

Weigelt, M. and Rumhor, H. (1986). Effects of wide ranging oxygen depletion on benthic fauna and demersal fish in Kiel Bay 1981–1983. *Meeresforschung* 31, 124–136.

Westermann, P. (1993). Wetland and swamp microbiology. In: Ford (1993), pp. 215–238.

Wetzel, A. (1929). Der Faulschlamm und seine ciliaten Leitformen. *Z. Morph. Ökol. Tiere* 13, 179–328.

Wetzel, R.G. (ed.) (1975). *Limnology.* W.B. Saunders, Philadelphia.

Whatley, F.R. (1981). The establishment of mitochondria: *Paracoccus* and *Rhodopseudomonas. Ann. N.Y. Acad. Sci.* 361, 330–340.

Whatley, J.M. and Chapman-Andresen, C. (1990). Phylum Karyoblastida. In: *Handbook of Protoctista* (ed. L. Margulis, J.O. Corliss, M. Melkonian and D.J. Chapman), pp. 167–185. Jones and Bartlett Publishers, Boston.

Whitman, W.B., Bowen, T.L. and Boone, D.R. (1992). The methanogenic bacteria. In: Balows *et al.* (1992), pp. 719–767.

Widdel, F. (1986). Suphate-reducing bacteria and their ecological niches. In: *Soc. Appl. Bact. Symp.* No. 13, pp. 157–184. Blackwell Scientific Publications, Oxford.

Widdel, F. (1988). Microbiology and ecology of sulfate- and sulfur-reducing bacteria. In: Zehnder (1988), pp. 469–585.

Widdel, F. and Hansen, T.A. (1992). The dissimilatory sulfate- and sulfur-reducing bacteria. In: Balows *et al.* (1992), pp. 583–624.

Widdel, F., Schnell, S., Heising, S., Ehrenreich, A., Assmus, B. and Schink, B. (1993). Ferrous iron oxidation by anoxygenic phototrophic bacteria. *Nature* 362, 834–836.

Wieser, W. and Kannwisher, J. (1961). Ecological and physiological studies on marine nematodes from a small salt marsh near Woods Hole, Massachusetts. *Limnol. Oceanogr.* 6, 262–270.

Wieser, W., Ott, J., Schiemer, F. and Gnaiger, F. (1974). An ecophysiological study of some meiofauna species inhabiting a sandy beach at Bermuda. *Mar. Biol.* 26, 235–248.

Wilde, P. and Berry, W.B.N. (1986). The role of oceanographic factors in the generation of global bio-events. In: *Lecture Notes in Earth Sciences, 8. Global Bio-Events* (ed. O. Walliser), pp. 75–91. Springer-Verlag, Berlin.

Williams, A.G. (1986). Rumen holotrich ciliate protozoa. *Microbiol. Rev.* 50, 25–49.

Williams, A.G. and Coleman, G.S. (1989). The rumen protozoa. In: *The Rumen Microbial Ecosystem* (ed. P.N. Hobson), pp. 77–128. Elsevier, London.

Williams, A.G. and Coleman, G.S. (1991). *The Rumen Protozoa.* Springer-Verlag, Berlin.

Williams, A.G. and Lloyd, D. (1993). Biological activities of symbiotic and parasitic protists in low oxygen environments. *Adv. Microb. Ecol.* 13, 211–262.

Williams, R.T. and Crawford, R.L. (1984). Methane production in Minnesota peatlands. *Appl. Environ. Microbiol.* 47, 1266–1271.

Winker, S. and Woese, C.R. (1991). A definition of the domains *Archaea, Bacteria* and *Eucarya* in terms of small subunit ribosomal RNA characteristics. *Syst. Appl. Microbiol.* 14, 305–310.

Woese, C.R. (1977). Endosymbionts and mitochondrial origins. *J. Mol. Evol.* 10, 93–96.

Woese, C.R. (1987). Bacterial evolution. *Microbiol. Rev.* 51, 221–271.

Woese, C.R. (1992). Prokaryote systematics: the evolution of a science. In: Balows *et al.* (1992), pp. 3–18.

Wolfe, R.S. (1992). Biochemistry of methanogenesis. *Biochem. Soc. Symp.* 58, 41–49.

Wolin, M.J. and Miller, T.L. (1987). Bioconversion of organic carbon to CH_4 and CO_2. *Geomicrobiol. J.* 5, 239–259.

Wolin, M.J. and Miller, T.L. (1989). Microbe-microbe interactions. In: *The Rumen Microbial Ecosystem* (ed. P.N. Dobson), pp. 343–359. Elsevier, London.

Wood, H.G. and Ljungdahl, L.G. (1991). Autotrophic character of the acetogenic bacteria. In: *Variations in Autotrophic Life* (ed. J.M. Shively and L.L. Barton), pp. 201–250. Academic Press, London.

Yamin, M.A. (1978). Axenic cultivation of the cellulolytic flagellate *Trichomitopsis termopsidis* (Cleveland) from the termite *Zootermopsis. J. Protozool.* 25, 535–538.

Yang, D., Oyaizu, Y., Oyaizu, H., Olsen, G.S. and Woese, C.R. (1985). Mitochondrial origins. *Proc. Natl. Acad. Sci. USA.* 82, 4443–4447.

Yarlett, N. (1994). Fermentation product generation in rumen phycomycetes. In: *Anaerobic Fungi* (ed. D.O. Mountfort and C.G. Orpin), Marcel Dekker, New York.

Yarlett, N., Lloyd, D. and Williams, A.G. (1985). Butyrate formation from glucose by the rumen protozoon *Dasytricha ruminantium. Biochem. J.* 228, 187–192.

Yarlett, N., Orpin, C.G., Munn, E.A., Yarlett, N.C. and Greenwood, C.A. (1986). Hydrogenosomes in the rumen fungus *Neocallimastix patriciarum. Biochem. J.* 236, 729–739.

Yarlett, N., Rowlands, C., Yarlett, N.C., Evans, J.C. and Lloyd, D. (1987). Respiration of the hydrogenosome-containing fungus *Neocallimastix patriciarum. Arch. Microbiol.* 148, 25–28.

Zebe, E. (1982). Anaerobic metabolism in *Upogebia pugeltensis* and *Callianassa californiensis* (Crustacea, Thalassinidea). *Comp. Biochem. Physiol.* 72 B, 613–617.

Zebe, E. (1991). Arthropods. In: Bryant (1991), pp. 218–237.

Zehnder, A.J.B. (ed.) (1988). *Biology of Anaerobic Microorganisms.* John Wiley and Sons, New York.

Zehnder, A.J.B. and Stumm, W. (1988). Geochemistry and biogeochemistry of anaerobic habitats. In: Zehnder (1988), pp. 1–38.

Zehnder, A.J.B. and Svensson, B.H. (1986). Life without oxygen: what can and what cannot? *Experientia* 42, 1197–1205.

Zehnder, A.J.B., Huser, B.A., Brock, T.D. and Wuhrmann, K. (1980). Characterization of an acetate-decarboxylating, non-hydrogen-oxidising methane bacterium. *Arch. Microbiol.* 124, 1–11.

Zhang, J.-Z and Millero, F.J. (1993). The chemistry of the anoxic waters in the Cariaco Trench. *Deep-Sea Res.* 40, 1023–1041.

Zierdt, C.H. (1986). Cytochrome-free mitochondria of an anaerobic protozoan — *Blastocystis hominis. J. Protozool.* 33, 67–69.

Zillig, W. (1987). Eukaryotic traits in archaebacteria. *Ann. N.Y. Acad. Sci. USA* 503, 78–82.

Zindel, U., Freudenberg, W., Rieth, M., Andressen, J.R, Schnell, J. and Widdel, F. (1988). *Eubacterium acidaminophilum* sp. nov., a versatile amino acid-degrading anaerobe producing or utilizing H_2 or formate. *Arch. Microbiol.* 150, 254–266.

Zinder, S.H. and Koch, M. (1984). Non-aceticlastic methanogenesis from acetate: acetate oxidation by a thermophilic syntrophic culture. *Arch. Microbiol.* 138, 263–272.

Zubkov, M.V., Sazhin, A.F. and Flint, M.V. (1992). The microplankton organisms at the oxic-anoxic interface in the pelagial of the Black Sea. *FEMS Microbiol. Ecol.* 101, 245–250.

Zumft, W.G. (1992). The denitrifying prokaryotes. In: Balows *et al.* (1992), pp. 554–582.

Zwart, K.B., Goosen, N.K., von Schijndel, M.W., Broers, C.A.M., Stumm, C.K. and Vogel, G.D. (1988). Cytochemical localization of hydrogenase activity in the anaerobic protozoa *Trichomonas vaginalis*, *Plagiopyla nasuta* and *Trimyema compressum*. *J. Gen. Microbiol.* 134, 2165–2170.

Index